Handbook of Next-Generation Emergency Services

Barbara Kemp
Bart Lovett

ARTECH
HOUSE
BOSTON | LONDON
artechhouse.com

Library of Congress Cataloging-in-Publication Data
A catalog record for this book is available from the U.S. Library of Congress.

British Library Cataloguing in Publication Data
A catalog record for this book is available from the British Library.

ISBN-13: 978-1-63081-652-0

Cover design by Mark Bergeron

685 Canton Street
Norwood, MA 02062

10 9 8 7 6 5 4 3 2 1

This book is dedicated to the 9-1-1 professionals that work tirelessly to save lives and sacrifice so much in terms of compensation and emotional capital. Their work often goes unrecognized and unappreciated by the general public. The people who answer 9-1-1 calls are literally the first responders to emergency incidents and almost never learn the outcome of these incidents or receive accolades for their actions. There are innumerable professionals that support these first responders both inside public safety agencies and outside the agencies as vendors and consultants. Without their collective dedication and keen minds none of it would be possible.

Contents

Acknowledgments

We wish to thank all those associated with this project: Artech House's Rachel Gibson, Emma Hobart, and David Michelson; NG-9-1-1, Inc. 9-1-1 SSP Team members Joe Cusimano, Rick Hird, Michael Ramsey, David Staub, and Travis Stender; Patrick Lustig, Ken Smith, and Melinda Woker at the Counties of Southern Illinois; Sandy Beitel and Glenna Johnson at the Northern Illinois Next Generation Alliance; Illinois Institute of Technology Professor Carol Davids, Texas A&M's Walt Magnussen, and Columbia University's Henning Schulzrinne and former FCC CTO, Solacom's Martin DeLeonardis and Ray Vilas, Oracle, Brian Kneuppel, Frequentis AG, Wolfgang Kampichlar, Frontier Communications and NENA Director Ronald Bloom, Neustar's Brian Rosen, Avaya's Mark Fletcher, Verizon's William Mertka, Mission Critical Partners' Marc Berryman, NENA Member Consultant Nate Wilcox, former NENA Operations Director Rick Jones, and Telcordia's Terry Reese. We would also like thank our families for their patience and support.

CHAPTER 1

Introduction to Emergency Services

This book aims to make the process of improving emergency services accessible to everyone regardless of background and experience. This chapter highlights emergency services from a historical perspective. The remainder of the book is designed to guide readers through planning, design, engineering, testing, conversion, implementation, and ongoing operations. Chapters are dedicated to legal, regulatory, and financial issues, while acknowledging the importance of security and privacy. A plethora of standards, guidelines, and processes support the narrative.

1.1 Emergency Services History

The desire for survival is embedded in the concept of obtaining rapid appropriate emergency services assistance. Emergencies happen to anyone, anywhere, at any time. Getting assistance is critical to saving lives and protecting property. The number 9-1-1 is the universal N11[1] number designated in North America to reach a first responder[2] for police, fire, and ambulance services.[3] 9-1-1 is an end-to-end service, and the technology that supports this service has been evolving for over 50 years, reaching a major turning point called Next-Generation 9-1-1 (NG9-1-1).[4] This book discusses how 9-1-1 technical solutions work and how to ensure that

---

1. N-1-1 is a number where the N can be any number from 1 through 9 followed by the digits 11. This is a 3-digit code assigned by the Federal Communications Commission (FCC) administered by the North American Numbering Plan Administration (NANPA)
2. The term first responder has been used to describe a number of individuals, from firefighters to Federal Emergency Management Agency (FEMA) personnel. However, with the introduction of the Middle-Class Tax Relief and Job Creation Act of 2012, a working definition needs to be agreed upon between federal and state agencies. The newly invigorated Middle-Class Tax Relief and Job Creation Act of 2014 provides grants to state and local governments for "emergency communications activities." However, the term 'first responder' isn't defined in the amended Job Creation Act ·PUBLISHED JANUARY 6, 2014 ·UPDATED MAY 11, 2016.
3. NANPA's responsibilities include assignment of NANP resources, and, in the United States and its territories, coordination of area code relief planning and collection of utilization and forecast data. NANPA assigns the N11 codes including 9-1-1. N11 codes are used to provide three-digit dialing access to special services. In the United States,, the FCC administers N11 codes. The FCC recognizes 211, 311, 511, 711, 811, and 911 as nationally assigned, but has not disturbed other traditional uses.
4. Next-generation 9-1-1 is a term coined by NENA, a standards creation body with headquarters in Washington, D.C., with an agenda to advance emergency services through political, legal, regulatory, and technical standards and initiatives.

this emerging emergency service platform meets caller[5] requirements by integrating with the aging public switched telephone network (PSTN) now and into the rapidly changing technological future. There were building blocks to move toward NG9-1-1. In sequence we address, operators, basic 9-1-1, enhanced 9-1-1 and NG9-1-1, discussing each topic in the context of the evolution of the technical capabilities that emerged to enable their success.

1.1.1 Telephone Operators—Pre-1968

Prior to the introduction of 9-1-1, emergency callers reached their local telephone company by dialing "0" for an operator or "4-1-1" for directory assistance if they could not dial directly. Operators requested the location of the caller and the nature of the emergency before contacting the appropriate police, fire departments, and ambulance services for dispatch. In some parts of the world today this is the full extent of emergency communications services. See Figure 1.1.

1.1.1.1 Operator Era Technology

The evolution of communications technologies and capabilities plays a large role in the introduction of network-enabled emergency services. Early telephone systems were electromechanical; most services were performed with the assistance of

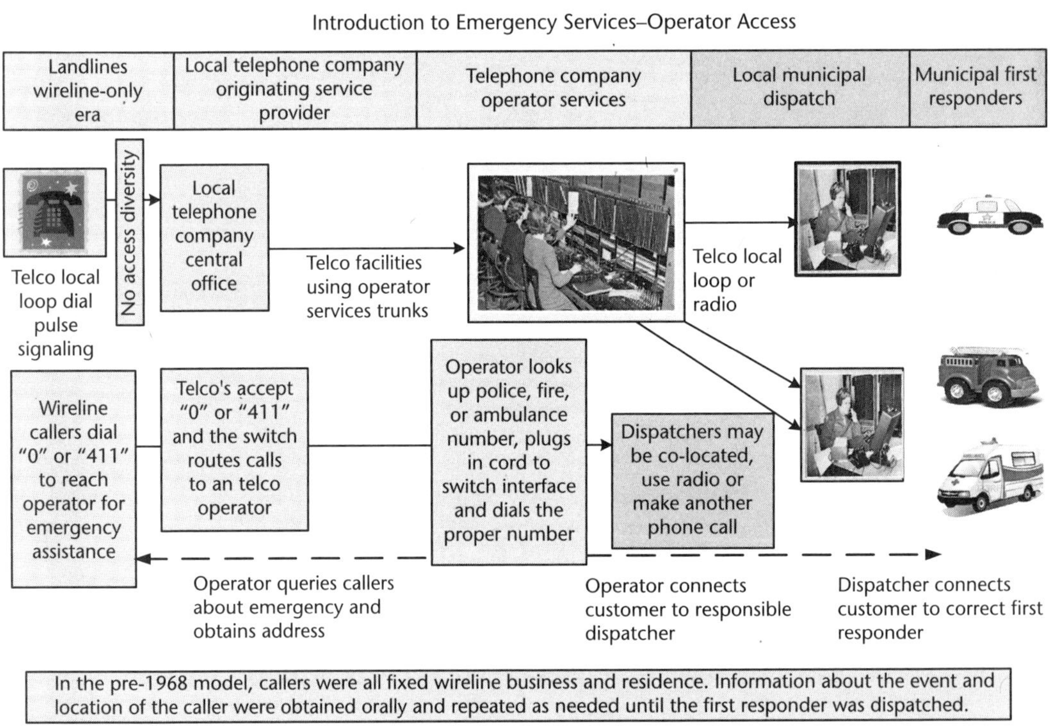

Figure 1.1 Operator.

5. Caller refers to the person making the request for emergency services regardless of the device used to make the call.

operators at switchboards making the connections from one caller to another and between telephone networks. The United States migrated to a seven-digit (7D) numbering scheme that included three-digit area codes, and local and long-distance services were defined. Trunk signaling paths between circuit switches predominantly used dial pulse[6] (DP) signaling. All telephone companies were monopolies. AT&T, the largest company, provided local and long-distance telephone service through 23 operating companies. There were approximately 1,400 regional independent telephone companies. The territories where telephone companies served were based on geography, and switches were deployed in so-called rate exchange areas. Emergency services were regulated primarily by state commissions with oversight by the Federal Communications Commission (FCC). AT&T's Bell Labs developed the standards for the United States and most of North America. In the 1960s, AT&T's goal was to complete the direct-distance dial[7] plan and attain universal service.[8] Telephone sets were hardwired and connected to telephone central offices (COs) via physical wires terminating to and cross-connected from telephone company frames to switches. We now refer to these original services as landline or wireline services. In this era, emergency services were all considered local. The local emergency providers are generally called public safety agencies. Public safety agencies used radio dispatch capability to contact police and fire departments. Assigned radio frequencies were not the same for all public safety answering locations, which eventually led to radio frequency complications when deploying 9-1-1 regionally and with statewide agencies. Public Safety Answering Points (PSAPs[9]) were limited

6. DP signaling is a signaling technology in telecommunications in which a direct-current local-loop circuit is interrupted according to a defined coding system for each signal transmitted, usually a digit. This leads to the often-used name loop-disconnect dialing. In the most common variant of pulse dialing, decadic dialing, each of the 10 Arabic numerals are encoded in a sequence of up to 10 pulses. The most common version decodes the digits 1 through 9, as one to nine pulses, respectively, and the digit 0 as t10 pulses. Historically, the most common device to produce such pulse trains is the rotary dial of the telephone, lending the technology another name, rotary dialing. The pulse repetition rate was historically determined based on the response time needed for electromechanical switching systems to operate reliably. Most telephone systems used the nominal rate of 10 pulses per second, but operator dialing within and between central offices often used pulse rates up to 20 per second. (Source: Wikipedia.)
7. Direct-distance dialing (DDD) is a telecommunication service feature in which a caller may, without operator assistance, call any other user outside the local calling area. Direct dialing by subscribers typically requires extra digits to be dialed as prefixes than for dialing within the local area or within an area code. also extends beyond the boundaries of the national public telephone network, in which case it is called international direct dialing or international direct-distance dialing (I). was the term used when the NANP was implemented in the 1950s. In the United Kingdom and other parts of the Commonwealth of Nations, the equivalent terms are or were "STD", for subscriber trunk dialing, and "ISD" for international subscriber trunk dialing. (Source: Wikipedia.)
8. Universal service is the principle that all Americans should have access to communications services. Universal service is also the name of a fund and the category of FCC programs and policies to implement this principle. Universal service is a cornerstone of the law that established the FCC, the Communications Act of 1934. Since that time, universal service policies have helped make telephone service ubiquitous, even in remote rural areas. Today, the FCC recognizes high-speed Internet as the 21st century's essential communications technology and is working to make broadband as ubiquitous as voice, while continuing to support voice service. (Source: FCC.Gov.)
9. PSAP is an entity responsible for receiving 9-1-1 calls and processing those calls according to a specific operational policy. Please note the following PSAP variations:
    - Primary PSAP: A PSAP to which 9-1-1 calls are routed directly from the 9-1-1 control office.

when trying to radio dispatch or transfer to neighboring agencies. Ultimately, the issue of public safety radio spectrum incompatibility came to a head in the United States during the New York City September 11, 2001, World Trade Center attack when many first responders were unable to communicate efficiently with each other contributing to a significant loss of life. Chapter 10 introduces solutions through FirstNet.[10]

In other countries, there has been a differing evolution of telecommunications, governed by a combination of private, government, and telecommunications entities using their own rules promulgated by international standards, with unique numbering schemes and systems. Chapter 13 addresses international emergency services.

1.1.2 Basic 9-1-1—1968

The first basic 9-1-1[11] system was installed in Haleyville, Alabama, in February 1968 to quickly connect a caller to the local police station without intervention from a telephone operator. Basic 9-1-1, an emergency telephone system that automatically connects 9-1-1 callers to a designated answering point, did not originally use automatic number identification (ANI) and never uses automatic location identification (ALI); call routing is determined by the originating CO switch only. The universal number 9-1-1 was introduced, and the first true PSAP emerged. Calls were routed to emergency services call takers,[12] interchangeably called telecommu-

---

- Secondary PSAP: A PSAP to which 9-1-1 calls are transferred from a primary PSAP.
- Alternate PSAP: A PSAP designated to receive calls when the primary PSAP is unable to do so.
- Consolidated PSAP: A facility where multiple public safety agencies choose to operate as a single 9-1-1 entity.
- Legacy PSAP: A PSAP that cannot process calls received via i3-defined call interfaces (IP-based calls) and still requires the use of CAMA or ISDN trunk technology for delivery of 9-1-1 emergency calls.
- Serving PSAP: The PSAP to which a call would normally be routed.
- NG9-1-1 PSAP: This term is used to denote a PSAP capable of processing calls and accessing data services as defined in NENA's i3 specification, NENA NENA-STA-010, and referred to therein as an "i3 PSAP".

REF: The NG9-1-1 definition was agreed to by NENA, NASNA, iCERT, and the National 9-1-1 Office representatives on 01/12/2018.

10. The FirstNet mission is to deploy, operate, maintain, and improve the first high-speed, nationwide wireless broadband network dedicated to public safety. This reliable, highly secure, interoperable, and innovative public safety communications platform will bring 21st century tools to public safety agencies and first responders, allowing them to get more information quickly and helping them to make faster and better decisions. See Firstnet.com.
11. Basic 9-1-1 is an emergency telephone system that automatically connects 9-1-1 callers to a designated answering point. Call routing is determined by originating CO only. Basic 9-1-1 may or may not support ANI and/or ALI.
12. Call takers are defined by NENA as telecommunicators who answer the calls in the PSAP, determining the location and nature of the emergency, and turn them over to dispatch or dispatchers and/or transfers the calls as required. Each call taker should have a personal login for security and be trained and cleared for the work they perform.

nicators, who would ask for clarifying information about the nature and location of the emergency and determine the correct Agency[13] to be dispatched. See Figure 1.2.

1.1.2.1 Basic 9-1-1 Era Technology

The National Emergency Number Association (NENA) defines basic 9-1-1 as follows: "an emergency telephone system which automatically connects 9-1-1 callers to a designated answering point. Call routing is determined by originating CO only. Basic 9-1-1 may or may not support ANI and/or ALI." Basic 9-1-1 came about during the emerging electronic and digital-switched technology era.[14] Local call origination migrated from the operator switching the call to a series of numbers that the circuit switches could interpret a digit at a time (i.e., 9-1-1) and determine the routing to the destination of the called party.[15] Basic 9-1-1 did not include the

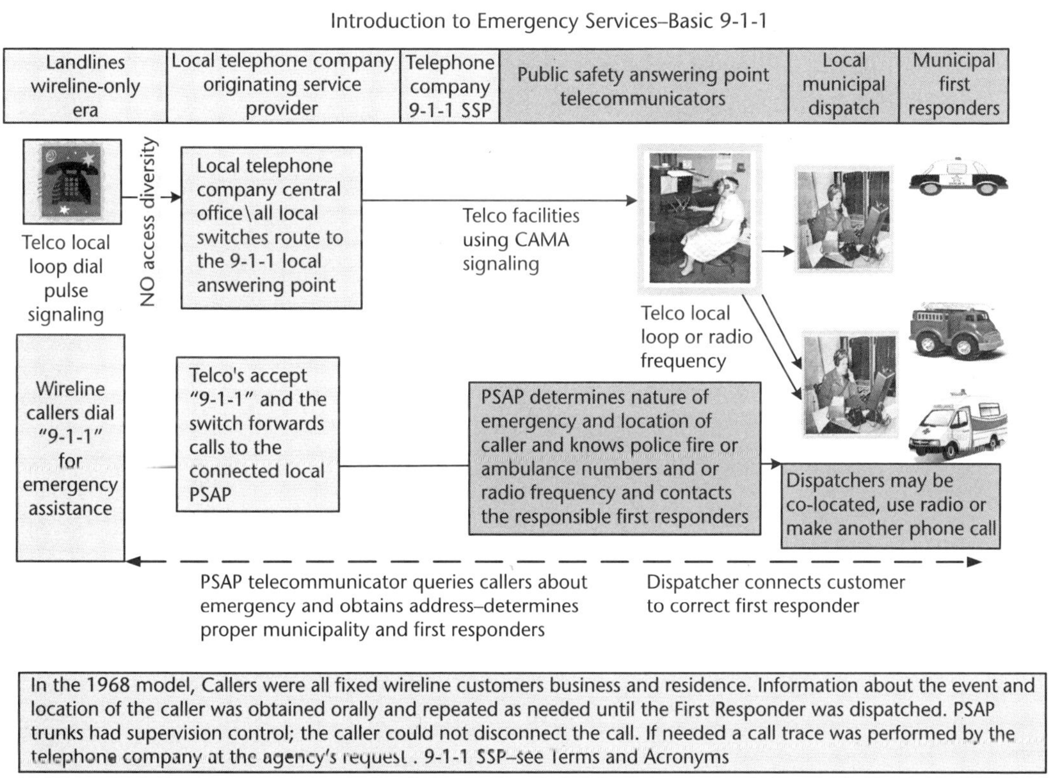

Figure 1.2 Basic 9-1-1.

---

13. Agency was initially the term referring to the various public safety organizations handling emergency call requests. The term has been refined and expanded for Next Generation 9-1-1 (NG9-1-1) to be more inclusive. Refer to Glossary of Terms and Acronyms.
14. Circuit and digital switches use a digital multiplexing technique for combining a number of signals into a single transmission facility by interweaving pieces from each source into separate time slots. Examples of circuit switches are ESS and DMS. ESS is a trademark of what is now Alcatel-Lucent; DMS is a trademark of what is now Gentech.
15. The called party (in some contexts called the "B-number") is a person who (or device that) answers a telephone call. The person who (or device that) initiates a telephone call is the calling party.

caller's location information (i.e., the address of the caller). The switch translations recognized the incoming party's request for emergency services and sent the call to the PSAP equipment and on to the call taker. The PSAP call taker interviewed the caller and extended the call to the proper dispatch agency or first responders by radio or by dialing the local seven-digit telephone number.

As basic 911 was deployed, local telephone companies opened[16] the code point 9-1-1 in each originating switch and routed calls over designated trunks, using DP signaling initially to the call taker who replaced the telephone company operator. The caller's originating local switch completed calls to the PSAP. The routing of the call from the caller's switch to the PSAP was done via centralized automatic message accounting (CAMA)[17] signaling trunks on dedicated trunk groups. Chapter 5 defines signaling options more comprehensively. CAMA technology used for 9-1-1 was initially designed to allow for a caller's telephone number identification to travel with the call to the telephone operator for long-distance billing *and* to allow the operator to "control" the caller's phone. CAMA trunks were repurposed for 9-1-1 because of those unique properties. CAMA trunks enable the PSAP call taker to control the call end-to-end, which means to control the physical path from the PSAP to the caller. If a caller tries to hang up, the call taker simply rings back the caller's handset or phone. The PSAP operator can contact the telephone company and request a "call trap and trace." A technician in the originating switch location traces the call to the caller's location, aided by customer records, to identify the physical address (i.e., location of the person making the emergency call). This function is embedded in law enforcement processes in U.S. Commission on Accreditation for Law Enforcement Agencies (CALEA)[18] standards. As to the basic 9-1-1-capable switch and location, the class-5 end-office switch was augmented with special functions to reach the dedicated local hardwired PSAP

1.1.3 Enhanced 9-1-1 (E9-1-1)[19]—1970s Forward

E9-1-1 "is a telephone system which includes network switching, database and PSAP premise elements capable of providing automatic location identification data, selective routing, selective transfer, fixed transfer, and a call back number". In Europe, a similar system exists, E1-1-2, which is known as eCall when used through a vehicle. An incoming call is selectively routed to the correct PSAP, and when the call is answered, the PSAP call taker's computer uses information provisioned through the telephone company's service order (SO) database to retrieve and display the

16. The term open herein refers to what was considered translations or wiring options using the numbering plan, in effect in each jurisdiction. A code went from vacant code to in-service and was turned up according to national and/or international guidelines.
17. CAMA is a special analog trunk originally developed for long-distance billing but is now mainly used for emergency call services: 911 and E-911. CAMA trunk connects a carrier switch directly to the SR, a special 911 switch that in turn connects to many PSAPs.
18. CALEA was created in 1979 as an accrediting authority through the joint efforts of law enforcement's major executive organizations. The CALEA accreditation program seals are reserved for the use of those public safety agencies that have demonstrated compliance with CALEA standards and have been awarded CALEA accreditation by the Commission. (www.CALEA.org.)
19. The term E9-1-1 also includes any enhanced 9-1-1 service so designated by the FCC in its Report and Order in WC Docket Nos. 04-36 and 05-196, or any successor proceeding.

physical address and/or the geographical coordinates of the caller. E9-1-1 was introduced in the 1970s because of the need to integrate the caller's location information into material to be displayed almost simultaneously with the emergency call. E9-1-1 was introduced in a wireline environment. Later with the introduction of wireless and voice over Internet Protocol (VoIP), there were additional modifications. In most of the world, E9-1-1 exists in one format or another. A pioneering E9-1-1 system was in place in Chicago by the mid 1970s, providing both police and fire departments access to the location of emergency calls. E9-1-1 networks are dependent on the legacy selective router (SR) tandem switches. Refer to Figure 1.3 for E9-1-1.

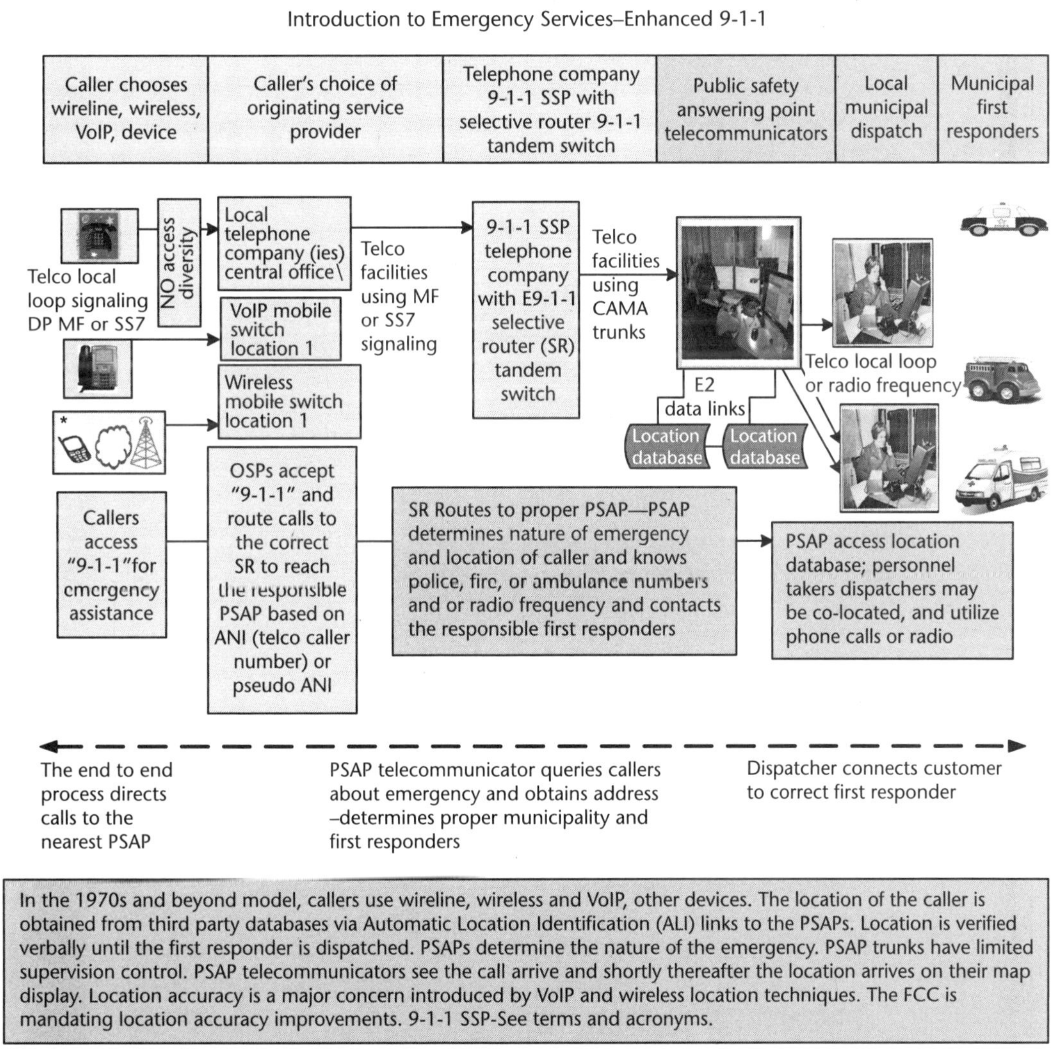

Figure 1.3 E9-1-1.

1.1.3.1 E9-1-1 Era Technology

Computing technology

The primary change with the E9-1-1 infrastructure was the introduction of computer terminals in the PSAP locations and the ability through computer technology integration (CTI)[20], for the call taker to retrieve key customer location information via specialized e2 datalinks, from diverse mainframe computers. The companies owning the mainframe computers store database records including selected portions of telephone company SO records (i.e., using standards-based formatted originating telephone numbers and the physical address of the caller location). Many third-party companies offer the database service, and individual telephone companies provide their own records systems and store the data in-house. Database management for location. Over time large telephone companies built their own ALI databases. When the call arrives at the E9-1-1 PSAP, from the incoming legacy SR, the CAMA trunks receive the ANI of the caller, and, through CTI, the PSAP performs a database lookup in real time to display the address of the caller to the PSAP call taker shortly after the call arrives. That retrievable address is loaded into the ALI databases after the customer places an order. The SO process was modified to get a service address (as opposed to billing address), loaded to the ALI database, so that when the E 9-1-1 calls arrive, the ANI is used to look up the location and display it to the PSAP call taker. Using this process, nearly all the wireline callers have a correct dispatchable address display in the PSAP even if the caller cannot speak. As with any computerized operations, updates can take from a day to several days before the caller's SO address is loaded to the ALI databases. Accordingly, a function was added to the telephone company responsibilities called the 9-1-1 database administrator. Each data field must be verified and entered in a NENA standard format. Rules were promulgated by state regulatory bodies for database accuracy, testing, and quality. The E9-1-1 requirements include computer-aided dispatch (CAD) systems and geographic information systems (GIS) database expertise.

Addressing and GIS technology

When GIS technology was introduced into the E9-1-1 systems in the 1980s and 1990s, it further refined location capabilities from callers with fixed-wireline technology. To make E9-1-1 a success, rural addresses were converted from locations such as intersections and rural route numbers to actual street addresses and from there converted to latitude and longitude or X, Y coordinates. GIS administrators are key to the success of E9-1-1.

Legacy SR—SR tandem technology

E9-1-1 was a significant upgrade from basic 9-1-1. For key detailed requirements, refer to NENA Generic E9-1-1 Requirements Technical Information Document, Issue 1, July 23, 2004. With the transition from basic 9-1-1 to E9-1-1 an important change was the addition of the legacy SR switch. The caller dials 9-1-1, and the

20. CTI is technology that connects phones to computers and allows them to interact. For agents, CTI puts data and call controls at their fingertips. It creates a link between the caller and information already in the customer relationship management (CRM) system, so they can be more efficient and provide better service.

phone company's originating switch, uses the caller's ANI capabilities and sends the call to the E9-1-1 SR switch, which acts as a hub (enhanced tandem) for the local 9-1-1 network. Originating access switches send their 9-1-1 calls over dedicated trunk groups to the local SR. The SR routes the calls to the appropriate PSAP.

Per NENA, "The Data Base Management System (DBMS) should be able to create selective routing Data Base (SRDB) data and deliver it electronically to either the switch (for those 9-1-1 SRs which use internal SRDB) or to the common pooled SRDB (for those switches that can go 'off switch' to query routing databases). The DBMS should ideally be able to update the SRDB throughout the day and not impact the SRDB query functions from the routing switch. The data should be secure." "The SRDB should respond to telephone number (TN) queries by the 9-1-1 SR switch and return the appropriate emergency service number (ESN). Where there is no ESN associated with a TN, then the SRDB should provide an ESN predicated on the programmed default logic. This can vary with each 9-1-1 SR's generic."

"The ALI database houses static location identification information for wireline callers and houses dynamic location information associated with wireless callers. When requested by the PSAP, the ALI database provides the existing location information for a caller to aid the PSAP agent in handling the call and to facilitate dispatch of emergency personnel to the caller's location. When the ALI database does not have the information requested by the PSAP but knows of another database that may contain this information, the ALI will route/steer the query to that database. In addition, the ALI database may need to coordinate the information from one or more databases in providing a response to the PSAP query." The PSAP uses the calling party ANI to access the address (location) of the caller from the ALI database and the emergency services–retrieved location address to determine the proper PSAP destination from the master street address guide (MSAG) This whole process is sometimes called selective routing, because in effect, the SR switch uses dynamic data to determine where to send the call instead of blindly routing it to a predetermined PSAP as in basic 9-1-1. In most cases, the location data retrieval step all takes a little over one second. Note that before the 9-1-1 call is made, the SO process diagrammed in the NENA standard describes the function of the SRDB.

Today the 9-1-1 tandem location identification reference exists in the local exchange routing guide (LERG) used by originating service providers (OSPs) to locate the proper 9-1-1 tandems (i.e., SR locations to route their call for emergency services).[21]

Signaling technology

Call signaling is a critical element in the migration to NG9-1-1 and should be considered by project managers as one of the highest priority critical path items for installation, testing, and deployment. Let's now describe the state of the technology.

21. In the LERG there is not a separate designation for Enhanced SR 9-1-1 Tandem, and the NG9-1-1 routing function designation needed for OSP interface. This book calls these tandems legacy SRs to distinguish them from the next-generation equipment that handles the Internet Protocol (IP) routing function. when IP evolved with E9-1-1,

SR signaling

CAMA signaling technology used between the Legacy SRs and the PSAPs is well over 50 years old; the test gear to test CAMA trunks is manufacturer-discontinued. Most legacy SR tandems are still in service using manufacturer-discontinued technology. This aging infrastructure is a key driver for migration to NG9-1-1. Look at Section 3.1.1 for Physical Network and the use of SPOF in that section. It is what it says in general terms, any single point of failure or a point that is not diverse. CAMA trunks are often a SPOF in the 9-1-1 call flow prior to NG9-1-1.

Originating signaling

Over time the caller's 9-1-1 originating signaling migrated from DP to predominantly multifrequency (MF)[22] signaling, then to Signaling System 7 (SS7) signaling starting with the first SS7 9-1-1 access location in Chicago Illinois in 1997. In the case of some large enterprises, the migration may include moving from MF to an integrated services digital network (ISDN)-primary rate interface (PRI)[23] where the legacy SRs have been modified to allow for that. Originating access signaling can be transitioned to ISDN-PRI, or in limited cases, to session-initiated protocol (SIP).[24] Chapter 5 describes OSP networks and signaling options, while Chapter 6 covers large enterprise direct connections and signaling. The book also addresses the migration to SIP signaling. This chapter defines SIP benefits for NG9-1-1.

---

22. MF is a signaling system that was introduced by the Bell System after World War II. It uses a combination of tones for address (phone number) and supervision signaling. The signaling is sent in-band over the same channel as the bearer channel used for voice traffic.
23. ISDN-PRI is a telecommunications interface standard used on an ISDN for carrying multiple DS0 voice and data transmissions between the network and a user. PRI is the standard for providing telecommunication services to enterprises and offices. It is based on T-carrier (T1) transmission in the United States, Canada, and Japan, while the E-carrier (E1) is common in Europe and Australia. The T1 line consists of 23 bearer (B) channels and one data (D) channel for control purposes, for a total bandwidth of 24 x 64-Kbps or 1.544 Mbps. The E1 carrier provides 30 B- and two D- channels for a bandwidth of 2.048 Mbps. The first timeslot on the E1 is used for synchronization purposes and is not considered to be a B- or D-channel. The D-channel typically uses timeslot 16 on an E1, while it is timeslot 24 for a T1. Fewer active bearer channels, sometimes called user channels, may be used in fractional T1 or E1 services. NG9-1-1 vendors do not currently provide connectivity options for fractional T1 or E1 services.
24. SIP is a signaling protocol used for initiating, maintaining, and terminating real-time sessions that include voice, video, and messaging applications. SIP is used for signaling and controlling multimedia communication sessions in applications of Internet telephony for voice and video calls, in private IP telephone systems, in instant messaging over IP networks as well as mobile phone calling over LTE (VoLTE).The protocol defines the specific format of messages exchanged and the sequence of communications for cooperation of the participants. SIP is a text-based protocol, incorporating many elements of the hypertext transfer protocol HTTP) and the simple mail transfer protocol (SMTP). A call established with SIP may consist of multiple media streams, but no separate streams are required for applications, such as text messaging, that exchange data as payload in the SIP message. SIP works in conjunction with several other protocols that specify and carry the session media. Media type and parameter negotiation and media setup is performed with the session description protocol (SDP), which is carried as payload in SIP messages. SIP is designed to be independent of the underlying transport layer protocol and can be used with the user datagram protocol (UDP), the transmission control protocol (TCP), and the stream control transmission protocol (SCTP). For the transmission of media streams (voice, video) SIP typically employs the RTP or the secure RTP (SRTP). For secure transmissions of SIP messages over insecure network links, the protocol may be encrypted with transport layer security (TLS).

Originating caller technology.
The advent of many technological changes, including, notably, the cellular phone, has shifted 9-1-1 demand. Today, approximately 10–20% of the calls for emergency services originate from traditional fixed wireline technology. The majority of 9-1-1 calls originate from wireless and VoIP[25] technology.

Wireless origination technology
Cellular telephones have brought significant changes and challenges to 9-1-1 service. Cellular telephones are categorized by regulatory bodies as commercial mobile radio service (CMRS); the 9-1-1 community has adopted the term wireless to describe the service. Wireless technology changed E9-1-1 requirements. First the caller devices are mobile, and the access network relies on a handset to tower relationship. The tower sends the calls using a mobile switching office (MSO) and trunks similar to the wireline trunks. Calls for emergency services are routed over trunks using SS7[26] to SRs the same as wireline calls. From there they are sometimes sent along a different CAMA trunk group to the appropriate PSAP. Separate wireline and wireless PSAP trunk groups are recommended because of the volume of wireless calls and the fact that many could be duplicated reports from the same incident. Using separate trunk groups allows the wireline (nonwireless) trunk group paths to remain available even if there is high demand for wireless emer gency callers. The goal is to keep wireless traffic from overloading wireline traffic.

Wireless database
NENA added requirements for E9-1-1 phase 1 and 2 wireless. Since a wireless telephone number can be used from any location in the world, the use of a fixed ANI number for caller identification could not reside in the wireline database. The new wireless database entries include the use of an area code and a fixed nondialable NXX code, using 555 for the first stage of wireless database identification. XXXX number ranges are assigned to each wireless OSP entering a market. NENA assigned each PSAP an ID digit number maintained in a national PSAP database. Wireless OSPs are required to notify a PSAP if they plan to enter their market, using the pseudocodes and database identification records, and to set up a testing plan. The 555 series numbers are known as pseudo ANI (pANI). When a mobile device dials 9-1-1, and the phase 1 wireless information pANI arrives in the PSAP; it shows that the associated call is coming from a wireless OSP and indicates the name of that OSP and the tower location. Phase 2 involves a connection to a mobile

25. VoIP: Technology that permits delivery of voice calls and other real-time multimedia sessions over IP networks. The Glossary of Terms and Acronyms more fully describes VoIP.

26. SS7 is a set of telephony signaling protocols developed in 1975 that is used to set up and tear down most of the world's PSTN telephone calls. It also performs number translation, local number portability, prepaid billing, short message service (SMS), and other mass market services.

In North America it is often referred to as the Common Channel Signaling System 7 (CCSS7). In the United Kingdom, it is called C7 (CCITT number 7), number 7 and CCIS7 (Common Channel Interoffice Signaling 7). In Germany, it is often called ZZK-7 (Zentraler ZeichengabeKanal Nummer7).

The only international SS7 protocol is defined by ITU-T's Q.700-series recommendations in 1988.Of the many national variants of the SS7 protocols, most are based on variants of the international protocol as standardized by ANSI and ETSI. National variants with striking characteristics are the Chinese and Japanese (TTC) national variants.

positioning center (MPC) entity that, working with the wireless OSP, will deliver the callers' location in X, Y GIS coordinates. In some cases, only the tower location is known. As of this writing, location accuracy is within an approximately 100-foot radius. The FCC requires continuous improvements in location accuracy.

VoIP origination technology

With the introduction of VoIP, there was a fresh need to look at E9-1-1. The technology was introduced with emergency services caveats from the OSPs. In the service contracts OSPs claimed that using their technology implied that 9-1-1 did not work from the VoIP telephone, computer, or device. Soon after VoIP introduction, it was clear that callers were using portable and fixed-VoIP technologies in locations where there was no alternative to reaching 9-1-1. Landlines for business and residences were being replaced. The FCC demanded that the VoIP providers add 9-1-1 access to their service offerings. Wireless providers, in many cases, use third-party providers to handle their 9-1-1 routing and database responsibilities. Third-party entities were expanding or formed for just that purpose. VoIP provisioning centers (VPCs) were established to help facilitate the FCC obligations. NENA adapted E9-1-1 standards to incorporate VoIP. By this time, there were more and more businesses that wanted to divest themselves of their role in 9-1-1 and continue with the more profitable parts of their businesses. Using the same model as wireless, VoIP OSPs were assigned pseudonumbers using the NXX code 222. The XXXX series was used to identify the VoIP OSP chosen by the caller. The VPC databases were established to assign the XXXX series to each VoIP OSP in a serving area. The VoIP OSPs notify each PSAP that they plan to connect and set up a test plan with the designated local telephone company SR tandem provider to connect their calls to the correct PSAPs.

VoIP database.

In lieu of the telephone company database, the records for the VoIP customers vary. VoIP devices can be fixed or mobile. The responsibility for the database records for mobile VoIP phones was left to the individual; in general, the caller must register the phone when moving from one location to another. This responsibility can mean that inaccurate location information reaches the PSAP and makes it especially difficult to dispatch if the caller either cannot speak or speaks a language other than the one familiar to the call taker. Fixed VoIP phones are the responsibility of the VoIP OSP to get the locations entered into the MPC or VPC databases for retrieval by the PSAP in near real time. Typically, E9-1-1 PSAPs have e2 datalinks to only one primary wireless or VoIP database company. The database providers exchange queries between themselves to retrieve the correct location information from the MPC or VPC provider. GIS locations are not always aligned with PSAP boundaries, and the callers can be moving while calling. PSAPs transfer callers to the correct PSAP, which takes time—and time is the enemy when handling emergencies.

VoIP callers were forced into an aging failing E9-1-1 paradigm. As a result, the goal for rapid and accurate emergency service response has been breaking at the seams. In response, the concept for NG9-1-1 emerged to rethink the process and redesign the systems end-to-end while maintaining service during the transition.

1.2 NG9-1-1

We turn our attention to today's opportunities and a revolution in emergency services, the focus of this book: NG9-1-1. "NG9-1-1 services includes a secure, Internet Protocol (IP)[27] based, open-standards system comprised of hardware, software, data, and operational policies and procedures that: a. provide standardized interfaces from emergency call and message services to support emergency communications; b. processes all types of emergency calls, including voice, text, data, and multimedia information; c. acquires and integrates additional emergency call data useful to call routing and handling; d. delivers the emergency calls, messages, and data to the appropriate PSAP and other appropriate emergency entities based on the location of the caller; e. supports data, video, and other communications needs for coordinated incident response and management; and f. interoperates with services and networks used by first responders to facilitate emergency response."[28] This book explains NG9-1-1 concepts and takes the reader through the steps to plan for, implement, and manage a NG9-1-1 program.

The book's 13 chapters culminate in challenges, plans, and opportunities for next-generation emergency services to be deployed anywhere in the world. A glossary of terms and acronyms is available for download. The NG9-1-1 vision is new; it is not a "bolt on" application of all that came before. NG9-1-1 must work seamlessly with the past and present, because it will take years to reach the vision. In the real world, we should not leave any caller behind. Standards are in a state of flux and in some cases simply incomplete. Experience with early adopters is crucial to the refinement of standards and implementation processes. Nothing is simple; next-generation emergency services technologies and services are being funded and rolled out in increments. Text to 9-1-1 is a good example, while the standards are in flux, vendors offer interim solutions that NENA, the wireless OSPs, and PSAPs have accepted to allow for this life-saving technology to meet the near-term demands from callers with disabilities. In one early state, wireless access to NG9-1-1-like early technical platforms were leveraged early because of a partnership with the state commission, which reduced costs and time to market and provided added advantages. Most of the time, instead of OSP switches connecting diversely and directly to the NG9-1-1 Emergency Services Internet Protocol Network[29] (ESInet) architecture with new next-generation core services (NGCS) equipment, the OSPs access the NG9-1-1 network via legacy technology. (Refer to Chapter 3 to learn

27. IP -The network in which the first publicly routable IP address is assigned to an endpoint. For residential broadband networks, the creation and supply of an access network may require the cooperation of several different providers. For example, an Internet service provider (ISP) may lease lines and digital subscriber loop access module (DSLAM) capacity from an existing telephony provider; in such a circumstance both entities are necessary to provide an access network.

28. REF: The NG9-1-1 definition was agreed to by NENA, NASNA, iCERT, and the National 9-1-1 Office representatives on 01/12/2018

29. An ESInet is a managed IP network that is used for emergency services communications and that can be shared by all public safety agencies. It provides the IP transport infrastructure upon which independent application platforms and core services can be deployed, including, but not restricted to, those necessary for providing NG9-1-1 services. ESInets may be constructed from a mix of dedicated and shared facilities. ESInets may be interconnected at local, regional, state, federal, national, and international levels to form an IP-based internetwork (network of networks). The term ESInet designates the network, not the services that ride on the network. See NG9-1-1 Core Services in the glossary available for download.

how NG9-1-1 callers predominantly use nondiverse routes and trunk groups to their legacy SR tandems; then calls are routed to the NG9-1-1 legacy gateway interfaces, and finally to the standards-based NGCS architecture and ESInet to complete. This is called a phased approach to NG9-1-1. Whichever method is chosen to begin the journey, the goal is to begin and complete steps as rapidly as possible to avoid managing and paying for two distinct environments, E9-1-1, and NG9-1-1. "As rapidly as possible" has taken on various meanings because of the intervention by commissions and legal intervention by the embedded 9-1-1 system service provider (911-SSP) responsible for the E9-1-1 network. A change to NG9-1-1 may or may not include a change in 9-1-1 SSP.

A few U.S. counties do not have basic 9-1-1 service, because they are too small or poor to buy and manage the equipment. According to industry experts, about 50 of the 3,100 counties in the United States do not have E9-1-1. The FCC requires OSPs to send 9-1-1-dialed calls to a defined answering point in those counties, by translating 9-1-1 to a seven- or 10-digit local sheriff office number, and not sending 9-1-1 to an announcement. Most OSPs in those counties have done that. Some regions that banded together to adopt NG9-1-1 brought underserved counties directly into their shared NG9-1-1 network. The remaining chapters in this book describe the necessary steps to achieve NG9-1-1 that includes the underserved communities. See Figure 1.4.

1.2.1.1 NG9-1-1 Era Technology

NG9-1-1 introduces new technology and databases. The bulk of NG9-1-1 standards were in draft form when early adopter locations began their journey toward NG9-1-1, pushed forward by their aging equipment and other societal and business factors. The dominant driver was to meet the needs of their local communities in serving their public safety mission. This book takes you through the experience of early adopters and updates readers on the status of the technology and its deployment in an ever-changing technically demanding environment. These are key enablers to allow for migration to NG9-1-1. Figure 1.4 represents the NGCS as one generic block. Chapter 3 defines the NG9-1-1 network detailing its internal components, and Figures 3.1–3.4 show each functional element (FE) that interfaces with the NGCS. Because of the legacy network and equipment, all interfaces include options to connect through legacy gateways.

SoftSwitch technology

The NENA standards definition of SoftSwitch technology is used here as quoted: "A software switch is a central device in a telecommunications network which connects telephone calls from one phone line to another, across a telecommunication network or the public Internet, entirely by means of software running on a general-purpose computer system. Most landline (Wireline) calls are routed by purpose-built electronic hardware; however, SoftSwitches using general purpose servers and VoIP technology are becoming more popular." Many telecommunications networks now make use of combinations of SoftSwitches and more traditional purpose-built hardware. "Although the term SoftSwitch technically refers to any such device, it is more conventionally applied to a device that handles IP-to-IP phone calls, while

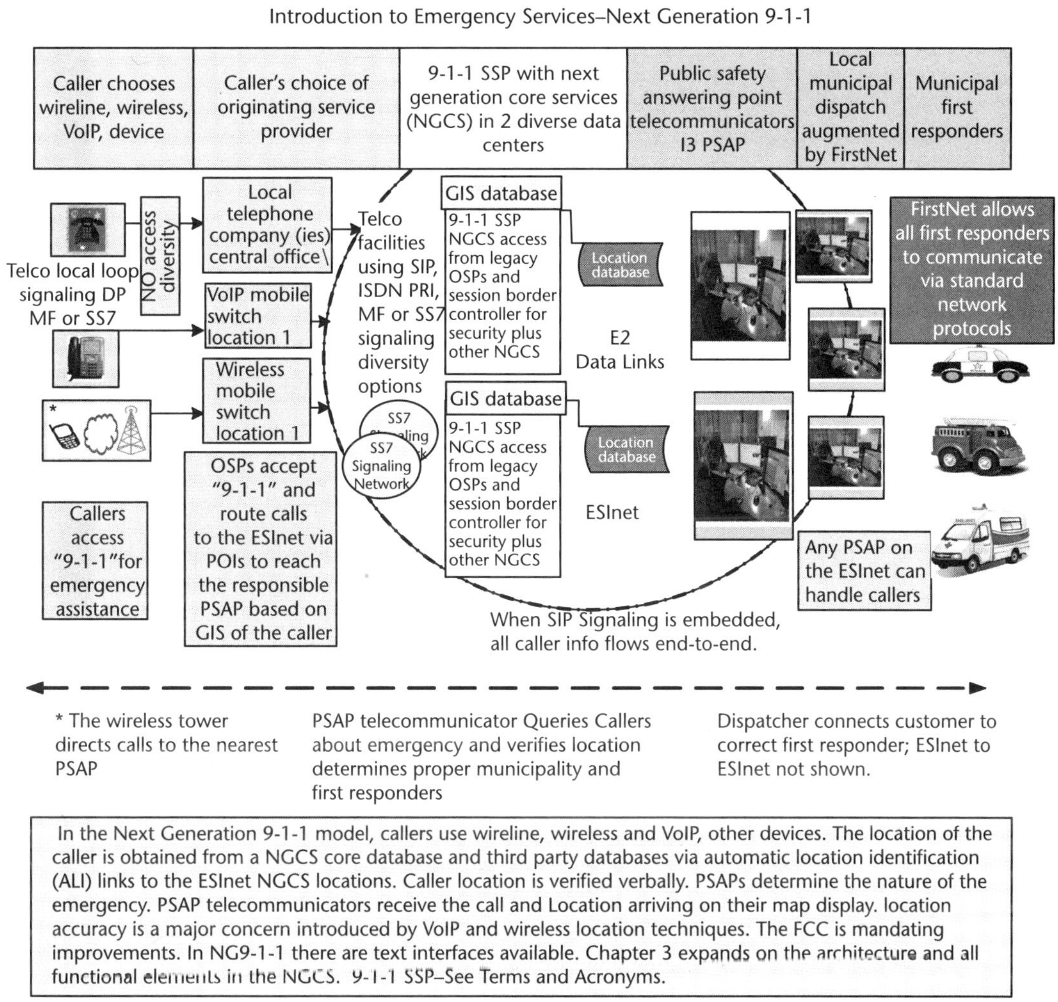

Figure 1.4 NG9-1-1.

the phrase 'access server' or 'media gateway' is used to refer to devices that either originate or terminate traditional 'landline' (hard wired) phone calls. In practice, such devices can often do both. As a practical distinction, a Skype-to-Skype phone call is entirely IP (internet) based, and so uses a SoftSwitch somewhere in the middle connecting the calling party with the called party. In contrast, access servers might take a mobile call or a call originating from a traditional phone line, convert it to IP traffic, then send it over the internet to another such device, which terminates the call by reversing the process and converting the Voice over IP call back to older circuit switched digital systems using traditional digital ISDN/PSTN protocols that transmit voice traffic using non-IP systems." The introduction of SoftSwitch technology enables changes from an end-to-end perspective with NG9-1-1. Few traditional landline OSPs have SIP signaling capability, but almost none use them in what is traditionally called a class 5 (telecommunications hierarchy) end office (EO) access environment. The first instantiation of SIP access for NG9-1-1 took place in the Counties of Southern Illinois (CSI) NG9-1-1 network when Clearwave, a class-5 OSP, offered 9-1-1 traffic to the CSI NG9-1-1 network from its Metaswitch across a diverse fiber-optic broadband network.[30] Clearwave's successful use of SIP was a milestone in 9-1-1 history.

SIP and real-time transport protocol (RTP) signaling

At the core of the NGCS in the NG9-1-1 network is the signaling known as SIP. In addition, any services that require real-time voice connectivity use RTP.[31] The benefits of these protocols is that SIP carries the caller's information along its path through the NG9-1-1 architecture, and that includes transfers between on-net next-generation PSAPs and off-net next-generation PSAPs reached through ESInet-to-ESInet connections. All the information pertinent to the caller and the calling location information is retained and does not have to be repeated or entered multiple times into new systems maps or databases.

Broadband network technology—the ESInet backbone of choice

While NENA standards do not dictate the network transport capabilities of the ESInet, they do mention that broadband fiber-optic networks have added bandwidth versus copper narrowband networks. ESInets can be sized to cover many areas including a county, region, state, or country; bandwidth is a major consideration in terms of engineering and capacity. A fiber-optic network configured into a true ring can provide added diversity and reduce the SPOF dominant in older networks serving basic 9-1-1 or E9-1-1 local areas.

Location-based routing in NGCS

The standards for NG9-1-1 database management are immense. While this book does not detail the database requirements, it does address the database at a high level, and explains how the access to the NG9-1-1 network location databases migrate and change. The wireline database migrates to a new diverse technology in the core of the NG9-1-1 architecture. Project managers should note that databases including a GIS entail one of the longest lead times in planning a NG9-1-1 system.

1.3 Summary

There is long history leading up to the requirements for building and managing a NG9-1-1 system. It is essential to integrate the NG9-1-1 capabilities with all existing technology and practices. A background in telephony is essential. 9-1-1 is first and foremost a service; the technology is secondary. The NG9-1-1 network reaches PSAPs that, in turn, must dispatch the appropriate agencies and first responders.

30. Metaswitch is a company that "develops software products and solutions that underpin the service platforms of today and tomorrow. Cloud-native virtual network functions are designed from the ground up to run on commodity hardware in private, public and hybrid clouds. This software powers the voice, data and unified communications services of fixed, mobile and converged operators worldwide, while also acting as the foundation on which incumbent operators and aggressive new entrants are building the cloud-based service platforms of the future."

31. RTP is an OSI model presentation layer protocol that provides end-to-end delivery of real-time audio and video by prioritizing this traffic ahead of connectionless data transfers. (See *ITP Principals and Applications* (p. 9) by D. M. Warts for more information.)

This chapter briefly addresses various means to access those first responders, including using devices such as telephones and radios. In some cases, the call takers or dispatchers text or email first responders, which is the last mile of information sharing. The goal is to take all that rich caller information and ensure that what is essential gets shared with the first responders onsite. The more the PSAP call taker must repeat or rephrase what is happening, the longer it takes to save a life or property. To have the same radio channel or communications devices across multiple agencies that respond to a fire or incident can save lives, including the lives of the first responders. This book's chapters build upon one another to simplify each step of the journey to success.

Selected Bibliography

NENA Generic E9-1-1 Requirements Technical Information Document Issue 1, July 23, 2004.

NENA-STA-010.2-2016 (originally 08-03).

CHAPTER 2

Anatomy of NG9-1-1

2.1 Project Scope

Major considerations for any NG9-1-1 project from concept to reality include project goals; geography served; public safety agencies and 9-1-1 SSPs operating within that geography; and technical, legal, and regulatory requirements . Countries and states may have a top-down approach. In parts of the United States, regional entities began the migration to NG9-1-1 on a pilot or trial basis. Leadership should be assigned primary responsibility to achieve the goals of NG9-1-1. NENA has state teams with designated leadership, using voluntary committee participation among members. The managers from the area(s) to be converted gather together to organize themselves into teams and create a request for proposal (RFP) to move forward. Subteams with specialized expertise support the goals. To get started there should be a planning team with a designated project leader, a project manager, an estimated budget, and target timeline. The planning team should dream big—and more specifically, review the current limitations of the area to be converted, document the triggers for change, understand industry requirements in governing regulatory documents and standards, and, where local changes are required, petition for changes. The planning team develops and releases the RFP, creates an award for the 9-1-1 SSP bidder's team, and proceeds to implementation.

2.1.1 Project Goals

Goals must be clearly defined. All agencies that serve the area to be converted must be identified. The planning team reviews the present method of operation (PMO) and the desired future method of operation (FMO) to identify gaps. When an area decides it is time to migrate to NG9-1-1, there are usually solid triggers for action since there is currently no law mandating the transition. Triggers may include, but are not limited to, frequent or significant outages; delayed response times; inaccurate or missing caller location identification; obsolete or manufacturer-discontinued equipment; aging telco infrastructure; inadequate bandwidth for newer types of data, video, or telematics; obsolete signaling; single points of failure; and/or staffing costs and expertise. Each agency has governing bodies that allocate funds for 9-1-1. Budgets may be squeezed so that agencies can no longer afford to maintain their own infrastructure. There may be a directive from a national or state level to provide better service for disabled citizens, to provide language services, or to increase reliability and accuracy for location services. By identifying the triggers and gaps, a solid plan emerges. Standards are available to identify how NG9-1-1 can fill the gaps. Armed with project goals, the process moves forward.

2.1.2 Geography

Public safety is delivered locally. The scope of a NG9-1-1 project may include areas as large as a country, a state/province, a region, or a county or as small as a single city, town, or village. We have each seen different forms of plan put into place; we are not advocating single-PSAP NG9-1-1 conversions. The NG9-1-1 standards allow for call takers to reside anywhere. Call takers can be located within a physical PSAP or work remotely as part of a virtual PSAP or through remote workstations. The key is to get the request for emergency assistance to the proper first responder—the one closest to the location needing assistance. The call must reach a PSAP call taker who can provide the most efficient service. NG9-1-1 represents a major change from how emergency services networks and technology were designed in the past. The ESInets and telco networks that support boundaries for the deployment of NG9-1-1 can be engineered to expand or contract to meet the demands of almost any geography. Public safety agencies and their PSAPs have geographical boundaries where they serve today. OSPs have boundaries where they serve today. There is no deliberate plan, agreement, or standards-based requirement to align the public safety agency or PSAP service boundaries with OSP service boundaries. It is important to match the callers and their OSP's switching locations to the public safety agency response. Experts in DBMSs[1] and GISs[2] are required to document the geography identified and create proper alignment. GIS and mapping data services[3] have long been the focus of the public safety agencies using road maps and jurisdictional boundaries, buildings, towers, parks, rivers, lakes, and any rational boundary to create an accurate mapping solution. There are many mapping systems companies, and as PSAP boundaries are crossed, there can be inconsistencies. Mobile and VoIP callers who need help may be on PSAP boundary borders and hard to locate. When NG9-1-1 is deployed, the mapping systems can be deployed across larger geographic areas, and map-to-map systems using NENA standards can more easily locate callers on the boundaries. One of the critical paths for any NG9-1-1 project is related to creating new standardized DBMS with standard mapping system interfaces so that calls from one PSAP area can be served by call takers in other PSAP areas and transferred seamlessly. Lives can be lost because of the lack of interagency interoperability. This all means that the geography to be served by

1. DBMS: A system of manual procedures and computer programs used to create, store, and update the data required to provide selective routing and/or ALI for E9-1-1 systems.
2. GIS: A system for capturing, storing, displaying, analyzing, and managing data and associated attributes that are spatially referenced.
3. Mapping data service: A service that returns images or features stored in a GIS that can be used to create a display for a telecommunicator or facilitate spatial analyses. Often used to provide maps for handling out of area calls, the mapping data service can also be used locally to provide a single, uniform map display for all functional elements in a PSAP that need maps.

the NG9-1-1 ESInet and network is another step to deciding how to organize, who is included, and the scope of the project.

2.1.3 Public Safety Agencies

Public safety agencies[4] have legal jurisdiction and responsibility for the citizens residing in and traveling within their boundaries. PSAPs exist with as few as two workstations to answer calls up to ever-increasing numbers of workstation sizes with dozens and up to hundreds of workstations. Some public safety agencies outsource call-taking responsibilities to neighboring jurisdictions. Other agencies have reciprocal agreements to serve as backup for neighboring jurisdictions (see Chapter 11). No matter how the citizens of a public safety agency are served today, there are a myriad of opportunities brought about by NG9-1-1 solutions. Whether driven by top-down or bottom-up approaches, public safety entities can invest together in technology and share common processes or best practices, personnel, and funding. Prior to NG9-1-1, each PSAP purchased its own technology to support its geographical footprint, installing its own information technology (IT) into PSAP telco closets, or police and/or fire department data centers, maintaining them from a privacy and security standpoint. Each agency staffed its own PSAPs. Most PSAPs had telephone company–dedicated personnel supporting them, connecting the PSAP networks with telco legacy SRs and assisting with engineering and planning decisions to size the transport facilities and PSAP equipment, including the work stations to meet peak demands. The NG9-1-1 standards move many of today's PSAP location-based FEs, which become part of the NGCS, into the new diverse NG9-1-1 qualified data centers[5] or network engineering building system (NEBS)[6]-compliant wire centers connected to a new ESInet. Next-generation PSAPs are called i3PSAPs.[7] Chapter 3 specifies what is in the core and what remains in the

4. Public safety agency and response agency are defined by NENA as having a legal or consensual obligation to respond to a call for service.
5. Data center: In terms of NG9-1-1, data centers are alternate locations to house the NGCS equipment. Data centers must have the same ability to host NGCS equipment as a NEBS-compliant CO. There is a strong recommendation that data centers be qualified as tier 3 or higher for NG9-1-1 networks. Contracts with SLAs are important. NGCS data center or CO locations must be approved by the state and/or oversight organization for NG9-1-1.
6. NEBS describes the environment of a typical U.S. regional Bell operating center (RBOC) CO. NEBS is the most common set of safety, spatial, and environmental design guidelines applied to telecommunications equipment in the United States. It is an industry requirement, but not a legal requirement. NEBS was developed by Bell Labs in the 1970s to standardize equipment that would be installed in a CO. The objective was to make it easier for a vendor to design equipment compatible with a typical RBOC CO. This would result in lower development costs and ease the equipment's introduction into the network. Telcordia now manages the NEBS specifications. The four then-largest U.S. telecommunications companies (AT&T, Verizon, BellSouth, and CenturyLink) created the Telecommunications Carrier Group (TCG), a group formed to synchronize NEBS standards across the industry in the United States. The TCG checklist specifies the individual NEBS requirements of each of its members in a matrix, making it simple to compare them. NEBS-complaint locations should meet level 3 standards. Wikipedia.
7. i3PSAP: This term is used to denote a PSAP capable of processing calls and accessing data services as defined in NENA's i3 specification, NENA NENA-STA-010, and referred to therein as an "i3 PSAP"., also known as NG9-1-1 PSAP. This book uses these terms interchangeably.

PSAP. The existing CAD[8] systems and capabilities tended to remain as they were aligned with each PSAP.[9] The number of remaining i3PSAPs and their staffing are part of the engineering and economic decisions for the public safety agencies. In the past, there was a focus on PSAP consolidation: call takers, gathered into brick and mortar buildings, to help the PSAP teams become more efficient. NG9-1-1 does not require that call takers reside in consolidated buildings. The NG9-1-1 NGCS are able to route calls anywhere along the ESInet to be answered as long as local first responder dispatch is possible.

2.1.4 Governance

Instead of local-only governance, a consolidated approach to governing, funding, staffing, vendor contracting, and providing IT support to the NG9-1-1 agencies is possible. Each public safety agency has its own hierarchical structure. Typically, there is a director in charge of a PSAP or group of PSAPs and that director reports to a government entity, such as a county commissioner or a police chief or mayor of a county, city or village or town with a board of trustees or alderman. Any and all decisions for expenditures require formal approval and legal contracts. Most entities must get competitive bids for any significant purchase. The decision to proceed to a NG9-1-1 project depends on RFP development, responses, analyses, recommendations, and approval. Chapter 12 addresses finances. Emergency services are also governed by federal government guidelines, rules, regulations, and depending on the country and state, guidelines, rules, and regulations. Public utility communications commissions have specific requirements for project approval, funding, and ongoing oversight. Chapter 11 addresses legal and regulatory matters. In North America, NENA[10] has developed evolving standards for NG9-1-1. State and regional NENA chapters adapt the standards and guidelines to fit what is needed at the local level. Not all governmental entities adopt all standards. International emergency services standards entities around the world either adopt or adapt the NENA or European Emergency Number Association (EENA)[11] standards to their country, laws, and requirements. The Association of Public-Safety Communications Officials (APCO)[12] is another public safety organization that works on similar issues, specifically radio and dispatch type issues. "APCO International was founded in 1935. It is the world's oldest and largest organization of public safety communications professionals and supports the largest U.S. membership base of any public

8. A CAD system is a real-time command-and-control process. These systems are used to track in real time all information relating to calls and units. These real-time command-and-control systems are often interfaced to E9-1-1 call systems to route calls between telephone call takers and dispatch operators.
9. Some state commissions have rules regarding numbers of PSAPs allowed to remain in service when converting to NG9-1-1. PSAP consolidation has long been a topic at NENA.
10. NENA is a not-for-profit corporation established in 1982 to further the goal of "One Nation-One Number." NENA: The 9-1-1 Association improves 9-1-1 through research, standards development, training, education, outreach, and advocacy. Our vision is a public made safer and more secure through universally available state-of-the-art 9-1-1 systems and trained 9-1-1 professionals.
11. EENA is a Brussels-based nongovernmental organization set up in 1999 dedicated to promoting high-quality emergency services reached by the number 112 throughout the European Union. (www.eena.org)
12. APCO is the world's oldest and largest not-for-profit professional organization dedicated to the enhancement of public safety communications. (http://www.apcointl.org/)

safety association." When moving toward NG9-1-1 in the United States, most of the standards documents come from NENA with APCO standards as needed. Once the NG9-1-1 project team goals are clear, and the standards referenced, there are other requirements for an RFP. In terms of governance, some states have established lead teams of experts to educate and define the goals for an RFP. The complexity and area to be converted may require added expertise. Table 2.1 lists the members of a sample RFP leadership team.

2.1.4.1 Rules and Regulations

Each state has emergency rules and regulations. If there is anything that prohibits working as a regional or local team or implementing NG9-1-1, those rules need to be reviewed for modifications or the project will not be successful. The FCC develops rules and regulations for wireline, wireless, and VoIP service quality. There are reporting rules for problems and notification to federal and state agencies that must be reviewed so a project can get approved. When the team documents its requirements, the regulatory rules that need to be enforced need to be included so that the winning 9-1-1 SSP commits to its plans.

2.1.4.2 Budget Management

Teams estimate their costs to manage and run dual networks while in transition, and they may need surcharge modification. Until the first bids come back, it is hard to budget for more than the existing operation plus the RFP and lead team management overhead. Assume that a new network and the existing network must run in parallel; then assume at least double current costs for a year for the network and operations. Some states have violated the public trust and allowed 9-1-1 funds to be diverted for other general needs. While this project is in effect, it may be wise to ask the governing entities to formally prohibit fund diversion. Strict project management, including financial accounting, must be in place for the entire project with jeopardy reports and escalation procedures submitted to the various boards and entities based on the requirements.

Table 2.1 Sample RFP Leadership Team

Title	*Voting Member Y/N*
Planning team lead	Y
Project manager	Y/N
Public safety managers 1-n	Y
Legal regulatory liaison	N
Database management leader	Y
GIS liaisons	N
Technical (IT, engineering) liaisons	N
RFP writer	N
Consultants (as required)	N

2.1.5 RFP Development

Funds are always scarce, and an RFP must be used to define the problem to be solved and request that bidders offer solutions based on standards and the needs of the communities that band together. The RFP needs to state clearly who is qualified to bid on a NG9-1-1 project. In many states the successful bidder must be a legal 9-1-1 SSP. Some entities require the successful bidder to be a certified local exchange carrier (CLEC) in the state it plans to serve. In most cases, the entity that develops the RFP cannot bid on the RFP. The clearer the requirements for the successful bidder are documented, the better the results. Lawyers must be involved each step of an RFP process to avoid conflicts and ensure fairness, preventing the bid from being invalidated and, thus, wasting time and money. The bidders may be required to respond with a systems integrator (SI)[13] providing a list of products and costs, delivery timetables, and ongoing operational plans. The RFP questions should follow the NENA NG9-1-1 standards and allow each respondent to answer them as yes or no and elaborate on differences. NENA standards provide many options. The RFP team needs to decide in advance on a particular option or let the respondents pick their choice and then evaluate which response best meets the goals after getting the bids. If the RFP team develops a list of its own requirements, how does it write the RFP or find a qualified RFP developer for NG9-1-1 to accomplish this process? An RFP can be prepared by the lead team or they can solicit and pay for help to prepare the RFP, get it sent, and make decisions. The planning team should not include any of the current 9-1-1 SSPs in the footprint on the lead RFP team; this causes a conflict of interest. Legal support is critical to an RFP process end-to-end. A final requirement is having a project manager to track the progress and budget and to ensure that reports are made and held confidential as required. Many documents will be developed and delivered, and proficient writing skills are required to assure that the work is tracked and accomplished professionally. Table 2.2 lists some sample RFP considerations in addition to the NENA standards.

2.1.5.1 Post-RFP Release

When it comes to the PSAPs, it is best that all RFP responders be given an opportunity to tour each PSAP that they will be serving both from a front-office perspective and from an IT interface perspective. There are always additional bidder questions. The procedure for questions and answers should be included in the timeline for the RFP process. Every bidder question needs to be seen by all bidders, along with all the answers. A meeting may need to be held before the first RFP release and then again with the first group of timely responders.

2.1.5.2 Evaluation of RFP

The planning team must decide if they want the RFP writer(s) to evaluate all responses with them or whether the lead team is qualified technically to manage this without additional expertise and expense. It is best to have a weighted numerical

13. The 9-1-1 SSP may hire a systems integrator to install and ensure interoperability of the network elements and the operations support systems. In some cases, the 9-1-1 SSP may be a systems integrator.

Table 2.2 Sample RFP Considerations in Addition to the NENA Standards

Questions by Category Requiring Short Answers with Descriptions in Reply
General
Does the bidder have reference clients?
Who are the major participants and what are their qualifications?
Database Management
Describe the database management of landline location services.
Describe the database management of wireless customer location services.
Describe the database management of VoIP location services.
Describe the database management of enterprise to NG9-1-1 location services.
Describe the support provided for the NG9-1-1 database management process.
Describe the GIS interface to DBMS.
ESInet
Describe and diagram the proposed ESInet.
Describe the design plan for ESInet's reliability, resiliency, and survivability.
Discuss environmental-, power-, grounding-, and fire suppression–related considerations for the ESInet.
Does the bidder provide the ESInet itself or under subcontract?
Is the ESInet fully diverse, and does the bidder require diverse components of the network for security, dynamic failover, and rerouting?
Data Centers or NEBS-Compliant Wire Centers
Where are the proposed data centers/wire centers inside the footprint?
What category are the data centers?
Do the data centers have their own network management systems?
How do the bidders handle the data center/wire center environment, power grounding, fire suppression, support for equipment location, and vendor access and wiring at POIs?
Do the bidders own data centers/wire centers or plan to contract with others?
What is the maintenance window for upgrades and work within the data centers?
NGCS or Core FEs
What vendor products are in their suite of NENA-compliant NGCS?
Do they insist on full duplication of FEs in the NG9-1-1 system?
How do they stress-test for load and overload before turning up for service?
PSAPs
Does the bidder recommend using E9-1-1 PSAPs or adding i3PSAPs from the beginning of the project?
How does the bidder manage physical and logical security for each PSAP?
How does the bidder handle environmental-, power-, grounding-, and fire-suppression–related situations at each PSAP?
How does the bidder perform training and documentation for each PSAP?
Does the bidder's proposal integrate with all unique PSAP systems and needs such as mapping, IP communications, dispatch systems, records management systems, logging recording systems, CAD, repair requests and notifications, alarm requirements, and user login tracking?
Testing
Does the 9-1-1 SSP have a laboratory for product integration? Describe the laboratory.
Has each product vendor been through the NENA ICE[1] testing? Describe each event and how participation was achieved in general (avoiding violation of confidentiality).
What is the bidder's testing process for the ESInet?
What is the bidder's testing process for the network to PSAP?
What is the bidder's testing process for the network with each type of OSP?
What is the bidder's testing process for each NGCS including the BCS and legacy gateways?
What is the bidder's testing process for all signaling, including SS7 and SIP?

Table 2.2 (continued)

What is the bidder's testing process for database and GIS boundary accuracy?
What is the proposed testing process for legacy network interfaces?
What is the proposed testing process for text to 9-1-1, data, and video?
What is the testing process for cutover?
Security
How does the bidder manage physical and logical security across the ESInet?
How does the bidder manage physical and logical security for each data center?
How does the bidder manage physical and logical security for each PSAP?
How does the bidder manage physical and logical security with enterprise customers connected directly to the NGCS?
How does the bidder manage physical and logical security end to end?
Where is the bidder's security system(s) located, and does the product comply with all NENA security standards[2] and NIST[3] standards?
OSPs
How do the bidders typically handle access integration from all OSPs?
Describe how the RFP is required to achieve diverse OSP access to the ESInet. Also, how does the bidder handle diversity, provisioning, testing, and cutover?
How does the bidder interface with MPCs?
How does the bidder interface with VPCs?
How does the bidder handle CLECs?
How does the bidder handle split exchanges?
How does the bidder manage the OSP interface, contracts, and SLAs?
Does the bidder expect to deploy in phases? If so, what shape will the phases take? When will the SPOF legacy SRs be retired?
Operations and Ongoing Management
How are the PSAPs able to log a request for assistance, and what response time is anticipated from the 9-1-1 SSP?
Does the bidder have local technical manpower within two hours of every PSAP and data center on the ESInet?
How does the bidder monitor and communicate information about service impacting conditions?
Does the bidder have a dedicated 9-1-1 network operations center (NOC)?
Where is the NOC and how is it staffed?
Is the RFP imposing service level agreements (SLAs) and penalties for nonperformance? If so, when does the bidder alert and how does it notify the Federal, State, and local public safety agencies of violations?
What participation is required of everyone in the testing process?
What participation is required of everyone in the cutover process?
What entity is ultimately responsible for service quality end-to-end in the eyes of the bidder?
Possible Additional Questions
Any requirements for location or ownership of the NGCS components should be specified.

1. NENA is dedicated to ensuring that the goals of NG9-1-1, both technical and nontechnical, are met. NENA views a range of testing programs as a critical component to accomplishing standards based NG9-1-1. NENA understands that it is the vendors of NG9-1-1 elements that will ultimately deliver the interoperability NG9-1-1 promises. NG9-1-1 testing events sponsored by NENA are aimed at supporting the vendors in their testing with each other. These industry collaboration events (ICEs) bring together vendors in an open, supportive, and collaborative environment that foster a spirit of technical cooperation. Taking part in or success in testing at an ICE event does not confer any formal NENA certification for a vendor's products.

2. NENA Standard 75-001, Security for Next-Generation 9-1-1 (NG-SEC). December 20, 2013.

3. National Institute of Standards and Technology (NIST) was founded in 1901 and is now part of the U.S. Department of Commerce. One goal of NIST is to provide the public safety community with a better understanding of what to expect from new and emerging networking technologies and accelerate the standardization and utilization of such technologies. NIST aims to "provide the public safety community with the performance analysis tools needed to better understand emerging network technologies and facilitate: the evaluation of worst/best case network deployment scenarios, the investigation of how well new technologies support public safety requirements, the development of quantitative requirements for public safety communications, the development of next generation network standards in support of public safety communication needs." This work is part of the Public Safety Communications Research program.

system for analyzing the replies and to use the same format and criteria for each responder. If an item has high, medium, or low importance to the planning team, it is helpful to the team and the respondent to identify them as such with the initial RFP.

When it gets down to the last qualified bidders, face-to-face demonstrations should be required. The 9-1-1 SSPs should take the planning team to their integration laboratory, or the planning team may ask them to set up a demonstration with presentations elsewhere, so it is easier to compare systems and solutions end-to-end.

2.1.5.3 Final RFP Selection and Contract

A formal process must be convened with a well-prepared report from the lead team on the winning bidder, describing any factors that disqualified the alternate bidders or that gave an advantage in a concrete, measurable way to the winner. Each entity may require a separate contract, or if a legal entity was formed with the public safety agencies involved, then a single contract may be possible. It takes time to provide a recommendation to every legal entity that may be involved, posting of meeting notices and presentations with follow up and final voting and award. An award does not usually guarantee that the actual plans will be approved for implementation. Based on discussions and the winning bidder's responses, budgets can be more clearly aligned for the final phases of the project, and a firmer timeline can emerge. Overall, project management tends to shift at this stage to a partnership between the 9-1-1 SSP and the lead planning team. Subsequently, schedules and the sequence of activities become clearer. It is seldom possible for all agencies to be cutover at once. Some planning teams ask their RFP team member(s) to step down, and others ask them to help with the management of the implementation, including overseeing documentation, testing plans in detail, and scheduling.

2.2 NG9-1-1 Project Management Teams

Once the contract has been awarded, there is a shift in overall project leadership from the entities to the 911 SSP or SI. Implementation teams are formed to address the most pressing needs for the geographic area and to find ways to move forward economically to implement the project. Team building, data gathering, and situational analysis are essential. Each implementation lead team member should have subteam support from the area to be served.

2.2.1 9-1-1 SSP Implementation Team

What kind of people do you need on your implementation team? Table 2.3 lists some possibilities. Please note that some members may wear multiple hats. Also, since voting is an issue within a specific area of expertise, it is necessary to decide who votes ahead of the first vote.

2.2.1.1 Added Considerations

An 9-1-1 SSP implementation team should also consider the following entities as members:

Table 2.3 Possible Implementation Team Members

Title	*Voting Y/N*
9-1-1 SSP Project Leader	
9-1-1 SSP Project Manager(s)	
Systems integrator	
NGCS vendor liaison(s)	
Statewide database Management liaison	
Database management Coordinators for the area to be converted	
GIS coordinator for the state	
GIS liaisons for area to be converted	
Network design engineer overall	
ESInet transport engineer	
IP engineer(s)	
9-1-1 telco engineer(s)	
Signaling engineer(s)	
Power environment Grounding, synchronization engineer(s)	
OSP leadership overall	
OSP coordinator—landline	
OSP coordinator VoIP	
OSP coordinator wireless	
OSP coordinator CLEC	
Enterprise coordinator for direct connections	
Testing coordinator(s)	
Legal regulatory liaison	
Commission liaison	
Client liaisons for each area to be converted	
Physical logical security expertise	
State or regional NENA director liaison to lead team	
State emergency services agency liaison(s) police fire EMS	
Legal and regulatory attorneys who understand telephony, tariffs, vendors, OSPs and NG9-1-1 concepts; Contracts extend to many areas. Refer to legal regulatory to determine if a single lawyer or multiple lawyers with expertise in 9-1-1 are required.	N
Consultants as required to fill gaps	N

- A nonvoting member OSP representative who is not an 9-1-1 SSP in the state, minimally one each from the wireline, wireless, and VoIP OSPs serving the area to be converted;
- An independent laboratory member, possibly a university member, who can assist with evaluating and obtaining laboratory support and testing;
- A nonbidding expert on ESInets and their design.

Ideally, the expertise of one individual may cover many of the topics. A state or country wants a smooth seamless NG9-1-1 experience. Assuming that is a larger goal, a planning team may wish to have statewide or country wide representation in important areas such as database management and GIS, because there are always boundaries and trying to get ESInet to ESInet integration ultimately depends on true or de facto standards. Given that NENA allows so many options, the standards interpretation may work well for a small ESInet, but it may not fit when trying to interoperate with a statewide ESInet or interstate ESInets. Accordingly, plan ahead and document decisions. Then, communicate them to all entities in the state or between states and countries even if they are not ready to deploy NG9-1-1 at the same time. Importantly, *every member should be subject to signed confidentiality agreements.*

2.2.2 Implementation Timeline.

The planning team along with the 9-1-1 SSP implementation team must establish a target for each major milestone in the NG9-1-1 implementation project. The 9-1-1 SSP project manager supporting the initiative must use formal processes and create an executive dashboard. (See Figures 2.1–2.3.) The dashboard should be updated and reviewed with the team weekly. Figure 2.1 presents a high-level overview and lists the project milestones and their status; it is not meant to capture every task but to set major milestones, assign primary ownership, and list the status and projected completion dates. Figure 2.2 lists outstanding items and due dates from various team members; these items should also be reviewed weekly to track progress on specific tasks. The Gantt chart outlined in Figure 2.3 is a visual tool to help the team see the relationship(s) between the different tasks or phases of the project in terms of duration and order to complete. Gantt charts can help the team understand the interdependencies and critical path of the project. This is not suggested as a substitute for more detailed planning and tracking. Rather, this approach can give the team an easy means to understand the situation and a consistent method of

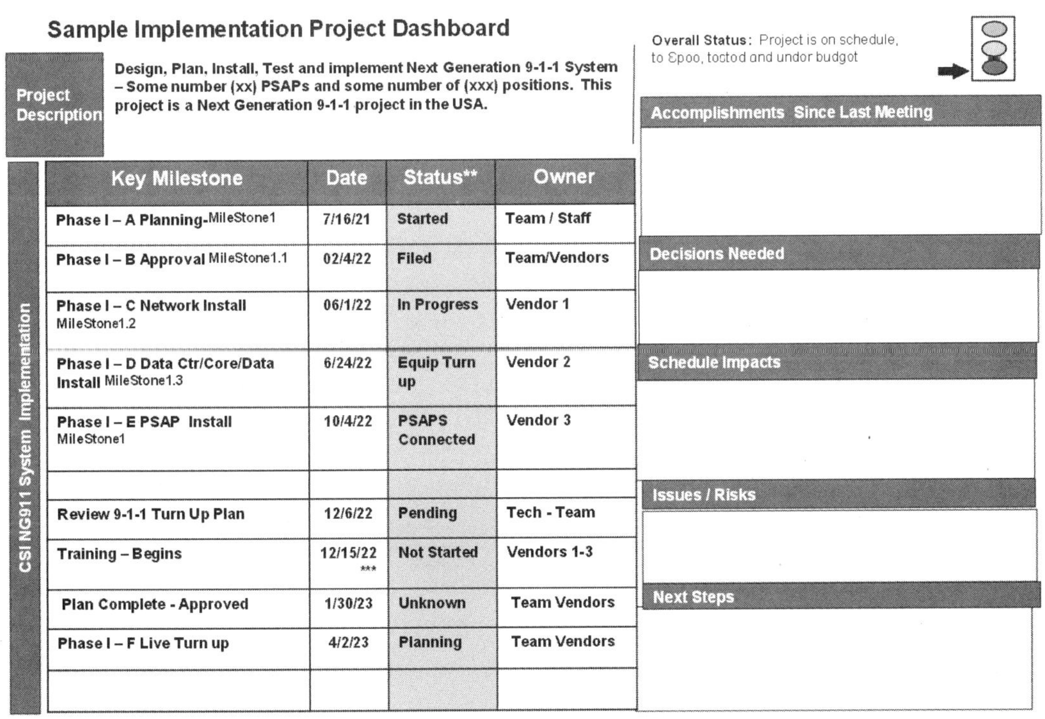

Figure 2.1 Time and overview.

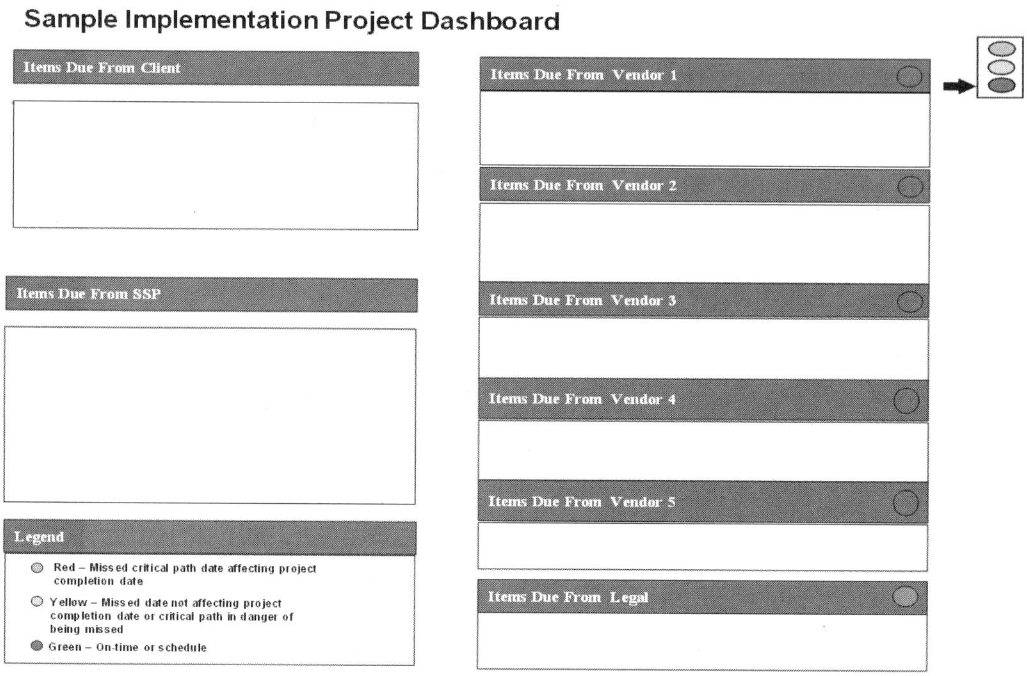

Figure 2.2 Action item register.

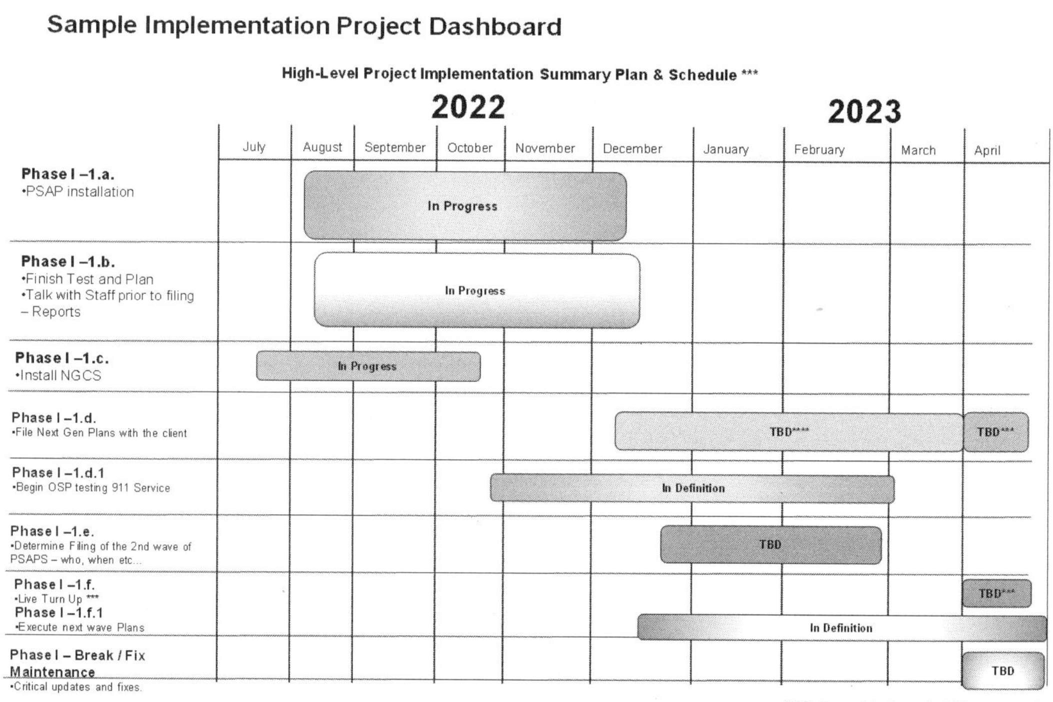

Figure 2.3 High level Gantt chart.

reporting. In every case there must be accountability and management of the SLAs in each vendor/carrier contract. The goal is to be successful by requiring bidders to adhere to the specified metrics and to provide an appeal process for exceptions and monetary penalties for nonconformance to the timing of the project plan.

2.2.3 Data Gathering

As part of the project, the historical information about the area to be served must be gathered. This includes each PSAP, each group of first responders (fire, police, and ambulance), and telephone companies and OSPs serving the area. It is also necessary to document all the existing GIS coordinators, database managers, and other vendors in the footprint and to take an inventory of the equipment and its physical status; existing IGAs need to be gathered and reviewed. The planning team must identify the attorneys who support each public service agency and determine who assumes leadership for efforts going forward in an integrated environment and how the group should reorganize and align itself legally. Other questions the planning team should consider are listed as follows:

- What name or moniker did the organization that released the RFP use (i.e., the name of the group or coalition)?
- Are there existing diagrams of the existing network, area codes, and local access transport area (LATA)[14] boundaries that are being impacted?
- What is the location of the existing legacy SRs in the footprint today and what is or are the existing 9-1-1 SSP(s)?
- How are calls currently transferred between agencies
- How many square miles or kilometers is each agency currently serving?
- What are the call volumes of each PSAP by date and time? Review the historical load and determine the busiest day, second busiest day, busiest hour, and second busiest hour.
- How many PSAPs are there, and how many positions are there per PSAP?

Finally, the planning team should agree on maintenance windows to be observed for service that could potentially affect work, and review how agencies work together now to determine what practices need to be reviewed for commonality.

2.2.4 Future State

The clearer the objectives of the project, the more likely it is that it will succeed. Once a bidder is chosen, contract negotiations can begin to complete the details. Regardless of the promises in the RFP response, a documented future state must be in place to complete the contract. No matter how much a group or entity may want a turnkey solution from the 9-1-1 SSP and SI, there are always key decisions. Consider the following examples:

14. LATA is a U.S. term that refers to a geographic region assigned to one or more telephone companies for providing communication services. A connection between two telephone companies within the same region is referred to as IntraLATA (Wikipedia).

- If the bidder promises to directly connect OSPs, will it be done in stages or at cutover?
- How long does the bidder have to connect the OSPs directly and remove the legacy SRs from the network? If the bidder offers a phased approach, what are the deadlines for each phase to be completed and what are the dependencies?

It is also necessary to consider that SLAs are dependent on actionable items. Therefore, planning teams must determine who pays what to whom if service is degraded during or after the cutover. While this may appear to be redundant with the RFP, often there are many details to be addressed such as priorities and sequencing of what agencies go first, second, and third. ESInets need to connect to other ESInets and/or back to legacy networks. ESInet and PSAP boundaries again come into the picture—and details are extremely important. Cooperation continues from start to finish. Accordingly, the following questions must also be answered:

- When is each PSAP ready to accept text to 9-1-1?
- Will full voice, video, and data come at the beginning of the turn-up for service, or will that be integrated and managed gradually?
- Will OSPs cutover one at a time by PSAP or by county or by state?
- Will the PSAPs get converted to i3PSAPs before or after NG9-1-1 network conversion?
- Will new intergovernmental agreements (IGAs) be needed ahead of time for call handling, or will the entities retain the same primary and back up arrangements?
- Will split exchanges be realigned?
- If consolidation occurs during the project, what are the surviving PSAPs, and where should OSPs route calls during the transition and afterward?

This is not a comprehensive list of questions; however, it allows the reader to springboard to other requirements. Training and documentation must be completed ahead of cutovers. PSAPs must be involved for testing and training. There are often local idiosyncrasies inside each PSAP or agency. If the agencies expect the 9-1-1 SSP to know about how they handle alarms when call takers take a break, then it needs to be documented. No detail is too small. Skilled 9-1-1 SSPs and SIs are familiar with this planning. Planning teams need to be in control of details.

2.2.5 Transition Planning

There are several things that need to be addressed when planning the transition to NG9-1-1. Chapter 3 addresses infrastructure. Be sure that the responders to the RFPs reference each item. Chapter 4 covers transfers to adjacent entities, discussing added steps that make the transition and transfers successful. Chapter 5 describes OSP access, while Chapter 6 considers large enterprise access. Chapters 7 and 8 detail testing and cutover planning, respectively. Chapter 9 discusses monitoring and network operations; Chapter 11 addresses legal regulatory issues; and Chapter 12 covers finances. Since this document is written at a "point in time," readers should always refer to the current standards. Chapter 10 addresses the roles that universities play in supporting NG9-1-1, allowing the planning team to understand possibilities and to be aware of pitfalls and rocky road scenarios traveled by the pioneers that made the journey to NG9-1-1. Many that offer NG9-1-1 service or claim to have implemented it are merely on the journey. Please determine which entities can really jump over the hurdles and get you to your vision quickly, economically, and safely. Once the planning and 9-1-1 system integration team has a fully documented and well-reasoned implementation plan, a major milestone involves gaining approval to proceed.

2.3 Application Approval

States often hold the keys to approval of any application for the transition to NG9-1-1. Therefore, it is necessary for planning teams to consult the state regulatory rules before beginning and to hold meetings to be sure that they fully understand governing requirements and are certain that the regulatory entities have a clear understanding of NG9-1-1. It is also necessary to build in time for the approval of plans; governing groups may be monitoring each interval, requiring reports. States often hold control over budgets for implementation and for ongoing funding. Accordingly, be aware of the hurdles and opportunities for good partnership. Assuming that you want to implement with text to 9-1-1, be aware there can be a six-month delay for the FCC application by the PSAPs and approval to proceed with that capability. We have seen years of delay when state 9-1-1 regulatory entities do not understand or agree with the plans.

2.4 Summary

NG9-1-1 must have a competent planning team and a documented plan to move successfully to approval and completion. This book is one guide to help planning teams navigate the plethora of documents and requirements to complete goals on time and within budget. The implementation team is equally crucial. This chapter helps readers to understand the roles between the teams and the importance of the close-knit relationship that must exist. A lack of transparency can lead to misunderstandings, which could cause a failure to gain approval and add significant cost and time.

Selected Bibliography

NENA-STA-010.2-2016 (originally 08-03)—NG9-1-1 System Architecture. NG9-1-1 Projects.

CHAPTER 3

Infrastructure and Database

3.1 Overview

NENA developed a set of standards for the successor design to E9-1-1. The standards are guidelines since there is no formal NENA oversight or penalty for noncompliance. NENA is in the process of developing compliance-testing requirements; it remains to be seen if there are consequences for noncompliance. As in many technical areas, a specific language is created with new terms and acronyms. This new language is helpful for professionals to communicate. At the same time, however, it can lead the casual observer to be confused or think the language is more complicated than it really is. This chapter demystifies the new language and sets the stage for the remainder of the book.

The largest change in NENA's successor solution to E9-1-1 is the transition to an IP-based design for the NG9-1-1 network. Transmitting voice calls over IP networks has been around since the late 1980s. The quality of a VoIP call has improved substantially since its inception and is being used in OSP networks to efficiently carry voice traffic. OSPs such as Vonage provide end-user voice telephone service solely using IP protocols.

Choosing IP for NG9-1-1 was a common-sense approach since most of the required hardware can be purchased off-the-shelf. There is nothing specific about how the "boxes" talk to each other; their means of communication did not need to be reinvented to handle NG9-1-1 calls. However, it was necessary to develop significant software, new databases, and the rules for handling the calls.

To keep NG9-1-1 secure and private, calls need to be routed over a private network. This means that the network that ties all the parts together is self-contained and has gateways to and from the network to control access in and out (i.e., ingress and egress). To distinguish this private network from other networks, NENA coined the term ESInet.

NENA members include the public safety community, manufacturers, OSPs, commonly called access carriers, consultants, and sales organizations for 9-1-1 equipment and systems. In more recent times, application (app) developers have joined NENA groups advocating for short-term fixes while waiting for standards to be thoroughly vetted—in some cases hoping their particular app will become a de facto standard. Chapter 10 addresses solutions beyond basic NG9-1-1.

Developing the NENA documentation is a voluntary committee process. Documentation that becomes standards provides options, and some parties advocate for solutions to be included in the requirements that are self-serving. NENA tries to limit self-serving solutions, but it is undeniable that certain members have more influence than others based on the size of their business, their financial contributions to conferences, and their ability to send paid members to participate in the voluntary industry process. It is advisable to review products developed according to NENA requirements with a critical design-oriented eye as opposed to choosing a hybrid solution with legacy components and/or easy solutions.

3.2 ESInet

An ESInet, a managed IP network that is used for emergency services communications, can be shared by all public safety agencies. It provides the IP transport infrastructure upon which independent application platforms and core services can be deployed, including, but not restricted to, those necessary for providing NG9-1-1 services. ESInets may be constructed from a mix of dedicated and shared facilities. ESInets may be interconnected at local, regional, state, federal, national, and international levels to form an IP-based internetwork (network of networks). The term ESInet designates the network, not the services that ride on the network.

The ESInet is ideally a privately managed IP network and does not operate across the open Internet. That does not mean that there cannot be limited access to the open Internet, but care must be taken on how to accomplish that interconnection while maintaining security and network reliability. It cannot be emphasized enough that the key to network reliability is redundancy and diversity. The ESInet uses IP protocol-based communications exclusively, and the bandwidth available on the ESInet determines much of the network's performance.

The NG9-1-1 system enables a wide spectrum of users to utilize emergency support services, with significant improvements for hearing-impaired, deaf, speech-impaired, and non-English-speaking callers through text to 9-1-1 and Teletype (TTY)[1] technology and through user identification and intelligent call-routing capabilities. The ESInet must be comprised of adequate bandwidth for voice, data, and video services. This requires significantly more bandwidth than traditional E9-1-1 networks. See Figure 3.1 for a sample ESInet configuration.

3.2.1 Physical Network

Physical transport is the baseline connecting and creating the NG9-1-1 system. The network may be built with copper cable, fiber-optic cable, or coax cable, or it may even involve wireless or satellite transmission. All of these can work, and with the

1. The phrase TTY (or teletype device) is how the deaf community used to refer to the extremely large machines they used to type messages back and forth over the phone lines. A TDD operates in a similar way but is a much smaller desktop machine. The deaf community has used the phrase "TTY," and sometimes uses it interchangeably with "TDD."

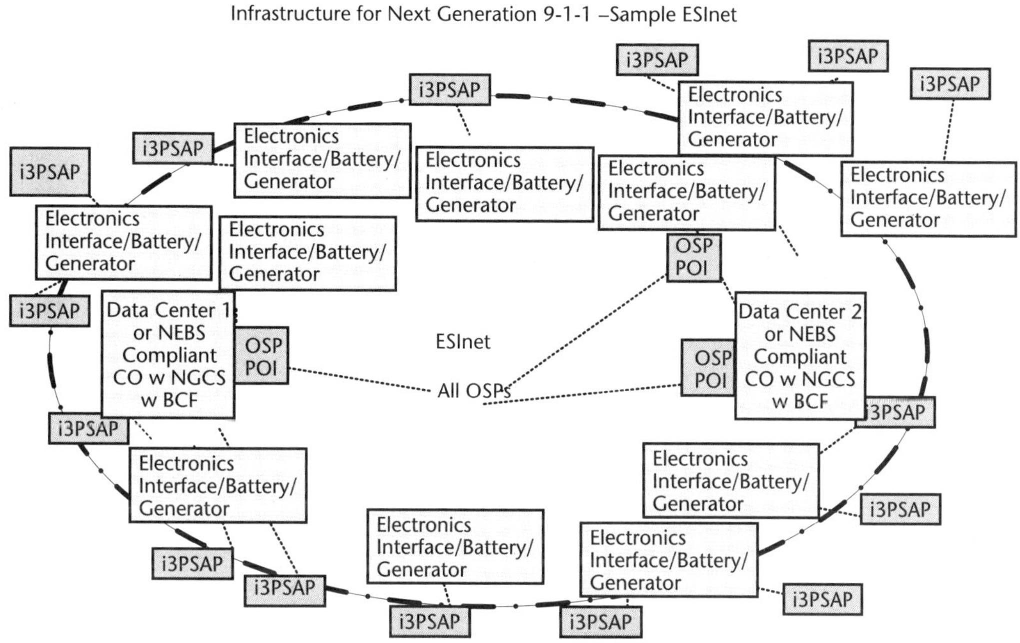

ESInet is comprised of 2 interconnected Fiber Rings with Electronics with Power Supplies Adequate for a minimum of 12 hour Outage Duration. There is one fiber string without diversity which often occurs. Rerouting of calls to an alternate PSAP is needed.	Next Generation Core Services (NGCS) in 2 diverse data centers or central office locations	i3 PSAPs only in this Example. All PSAPs and data centers connected on the ring. Additional equipment allows for full diversity. If failure in one part or component on the ring, traffic should be dynamically rerouted to complete through alternate NGCS component or fiber optic technology. POI—Point of Interface

Figure 3.1 Sample ESInet configuration.

proper engineering, they can be intermixed to create a fault-tolerant and robust ESInet to connect the next-generation FEs using IP signaling. Refer to the open systems interconnection (OSI)[2] model for the official layers of the network.

The design of the network that connects all the components of the NG9-1-1 system is critical. The choice of transport is the baseline for the reliability, resilience, and redundancy critical to the success of project deployment. Ideally the network chosen is a physical fiber ring with the NGCS and i3PSAPs residing on the ring, and not a drop off the fiber ring, which can introduce another SPOF. The availability of locations being physically on the ring may be limited due to location and cost. Economic and engineering considerations and tradeoffs occur when designing an ESInet.

2. OSI model: A conceptual model that characterizes and standardizes the communication functions of a telecommunication or computing system without regard to its underlying internal structure and technology. Its goal is the interoperability of diverse communication systems with standard protocols. The model partitions a communication system into abstraction layers. The original version of the model defined seven layers: application (layer 7); presentation (layer 6); session (layer 5); transport (layer 4); network (layer 3); data link (layer 2); and physical (layer 1).

3.2.2 Transport Options

The ESInet backbone includes some type of telecommunications or data circuits connecting the FEs. As a practical guide, nothing below a T1[3] (1.544-Kbps[4]) data rate should be used. In fact, the NGCS interfaces for existing NG9-1-1 vendor products require interfaces of T1 or above.[5] SIP voice calls alone consume somewhere from 12 to 96 Kbps per call. It is important to remember that more than the SIP call is traversing the network. There is map data and CAD data in addition to the voice, video, and data traffic. Ideally the ESInet should be built on a fiber-optic cable ring that can deliver from 100 Mbps[6] to 100 Gbps.[7] The ring provides reliability in that it is designed so that if one part of the ring is cut or electronics are inoperable, the traffic dynamically moves in the other direction, thus preventing a SPOF.

The next layer of the network contains a mixture of circuits. T1s or DS3s[8] could be connected to a fiber ring,[9] along with metro Ethernet (Metro-E)[10] or multiprotocol label switching (MPLS)[11] circuits.

---

3. The T1 carrier is the most commonly used digital transmission service in the United States, Canada, and Japan. In these countries, it consists of 24 separate channels using pulse-code modulation (PCM) signals with TDM at an overall rate of 1.544 Mbps. T1 lines originally used copper wire but now also include optical and wireless media. A T1 outstate system has been developed for longer distances between cities. It is common for an Internet access provider to be connected to the Internet as a point-of-presence (POP) on a T1 line owned by a major telephone network. Many businesses also use T1 lines to connect to an Internet access provider.
4. Kbps: One kilobit per second equals 1,000 bits per second. This is sometimes written as kbps, Kb/sec, or Kb/s, but all of them carry the same meaning.
5. International Standards often require E1 interfaces. The E-carrier is a member of the series of carrier systems developed for the digital transmission of many simultaneous telephone calls by time-division multiplexing. The European Conference of Postal and Telecommunications Administrations (CEPT) originally standardized the E-carrier system, which revised and improved the earlier American T-carrier technology, and this has now been adopted by the International Telecommunication Union Telecommunication Standardization Sector (ITU-T). It was widely adopted in almost all countries outside the United States, Canada, and Japan. E-carrier deployments have steadily been replaced by Ethernet as telecommunication networks transition toward all IP.
6. Mbps, or megabits per second, refers to download and upload speeds. However, there is another very similar acronym out there that mean a different thing: MBps, which means megabytes per second.
7. Gbps stands for one gigabit per second and equals 1,000 Mbps, one million Kbps, or one billion bps.
8. Digital signal level 3 (DS3) service consists of a high-capacity channel provisioned for transmission speeds of 44.736-Mbps isochronous serial data. A DS3 facility can be channelized to provide 28 DS1 circuits with the multiplexing ability to enable a platform for voice, video, or data services
9. Fiber optics are an optimal means to acquire adequate bandwidth for NG9-1-1 network deployments. The 9-1-1 SSP has choices for ESInet infrastructure. Fiber rings provide optimal performance for failover to achieve "five nines reliability."
10. Metro-E, a metropolitan area network (MAN), based on Ethernet standards, is commonly used to connect subscribers to a larger service network or the Internet. Businesses can also use metropolitan-area Ethernet to connect their own offices to each other.
11. MPLS is a type of data-carrying technique for high-performance telecommunications networks that directs data from one network node to the next based on short path labels rather than long network addresses, avoiding complex lookups in a routing table. This mechanism allows network administrators to perform a measure of traffic engineering within their networks.

3.2.3 Diversity and Capacity

To achieve high availability, reliability, and resiliency in the NG9-1-1 network, redundancy is critical. Murphy's Law teaches us that it is not a matter of whether failures will occur but when. There are multiple components that comprise the ESInet and NGCS. Failures can happen in any of the components from the base network, to the facilities, switches, routers, databases, workstations, gateways, and more. Redundancy in the NG9-1-1 network end-to-end is critical. Redundancy should be implemented from the call-origination point, the OSPs, through to the databases, the PSAPs, switches and routers, and to some degree workstations where calls complete.

Geographic diversity is also critical for locating the FEs of NG9-1-1 along the ESInet to minimize weather events and natural disasters having a negative impact on multiple locations if they are located closely together. Data centers should ideally be 50 miles or more apart. In some cases, ESInets cover multiple states. Significant service-impacting outages have occurred impacting NG9-1-1 services.

NG9-1-1 networks can be built with copper using T1 or even DS3, and they can reside on SONET[12] networks; these are aging technologies that are not as well-suited for the demands of a NG9-1-1 network let alone the cost for redundant facilities. Metro-E and MPLS are newer technologies and can be used over fiber.

3.3 Data Centers

Data centers, here used as a "term of art," are locations that house the NGCS. Data center locations should be geographically diverse, fully redundant, environmentally controlled, monitored, and highly secure. Data centers should be located in buildings on the ESInet.

NG9-1-1 data center locations housing the fully duplicated NGCS equipment must have the same technical ability to host NGCS equipment as a NEBS-compliant telephone company wire center. We strongly recommend that data centers be qualified as tier 3[13] or higher for NG9-1-1 networks. Contracts with SLAs are part of legal agreements, and locations must be generally approved and enforced by a governmental oversight organization for 9-1-1 service.

3.3.1 Data Center Ratings[14]

3.3.1.1 Tier 1 Data Center

A Tier 1 data center maintains a single path for all capacity loads. This type of data center has no redundant paths or fault-tolerant infrastructure. These data centers are particularly vulnerable during an incident. This type of data center, which

12. Synchronous optical networking (SONET) and synchronous digital hierarchy (SDH) are standardized protocols that transfer multiple digital bit streams synchronously over optical fiber using lasers or highly coherent light from light-emitting diodes (LEDs). At low transmission rates data can also be transferred via an electrical interface.
13. Refer to the Glossary of Terms and Acronyms for a description of Tier 1–4 data centers.
14. Source for data center rating definitions was an Internet search. Refer to the Glossary of Terms and Acronyms.

should be evaluated in terms of potential downtime per year, is not fault-tolerant or decentralized and has single distribution paths, limited continuous heating and cooling during a utility outage, and limited transition to emergency generator. This means that they are inadequate for NG9-1-1.

3.3.1.2 Tier 2 Data Center

A Tier 2 data center can support the capacity load across multiple connections but lacks redundant paths and fault tolerance. While it can remain functional for a period with some level of service degradation, it has no fault tolerance and has heating and cooling issues and limited emergency generators. This means that they are inadequate for NG9-1-1.

3.3.1.3 Tier 3 Data Center

A Tier 3 data center can support the capacity load across multiple connections and does contain an additional level of redundancy. It includes alternate distribution paths, fault tolerance, decentralization, limited continuous heating and cooling during a utility outage, and limited transition to emergency generators. Tier 3 data centers prioritize physical and cybersecurity and are monitored and staffed adequately. With multiple duplicated data centers, this type is adequate for NG9-1-1.

3.3.1.4 Tier 4 Data Center

A Tier 4 data center can support an outage using multiple connections. In addition, this type of data center can support the capacity load during a failure without limitation, simultaneously operating alternate distribution paths. They are also fault-tolerant and decentralized and offer continuous heating and cooling during a utility outage, immediate transition to an emergency generator, and full physical and cybersecurity protection. With multiple duplicated data centers, this type is adequate for NG9-1-1.

The data centers or telephone company NEBS-compliant wire center location, contracts must be stringent and include nonperformance penalties in SLAs Private data centers are often able to deny or disconnect their clients' services for nonperformance; that is not acceptable for NG9-1-1 SSP clients. If a data center goes out of business, the lead time for notification to the 9-1-1 SSP is critical. Chapter 11 addresses legal and regulatory matters.

It is critical to have a minimum of two equally equipped data centers for a legitimate NG9-1-1 network. Certainly, there have been installations that do not meet these criteria; yet, they call themselves NG9-1-1 because they purchased the technology with that name.

Points of interface (POIs) reside in each data center to facilitate termination of access equipment coming in from OSPs and termination of interfaces out to each PSAP or other 9-1-1 SSP network ESInets. There must be technical capabilities to cross-connect facilities from the POIs to the NGCS. Typically, non-telco 9-1-1 SSPs do not provide digital cross-connect systems. This can be a thorny matter if the responsibility for connectivity is not clarified in contract negotiations. Terminations vary depending on the total bandwidth required and the access diversity chosen by each OSP. The 9-1-1 SSP is responsible for the security component, the border control function (BCS), located in each data center, protecting the core of the NGCS, the PSAPs, and ESInet-to-ESInet connectivity, all interfaces to other entities. See Section 3.4.1 for BCF definitions. Figure 3.2 shows a single data center example. Data centers are not comprehensively addressed here. Legal negotiations and technical requirements are the responsibility of all parties and can keep a plan from being approved if not in alignment with the regulatory agencies' requirements.

3.4 NGCSs

The theoretical design model for NG9-1-1 breaks the various functions that must be performed to accept, identify, route, display, and record call information to call takers into the correct PSAP. NGCSs are comprised of and FEs that are contained in each data center. Each data center should have a fully duplicated set of NG9-1-1 equipment that is the "core" of the system. Section 3.4.1 through 3.4.17 describes the FEs for each data center. Figure 3.3 shows the NGCS architecture.

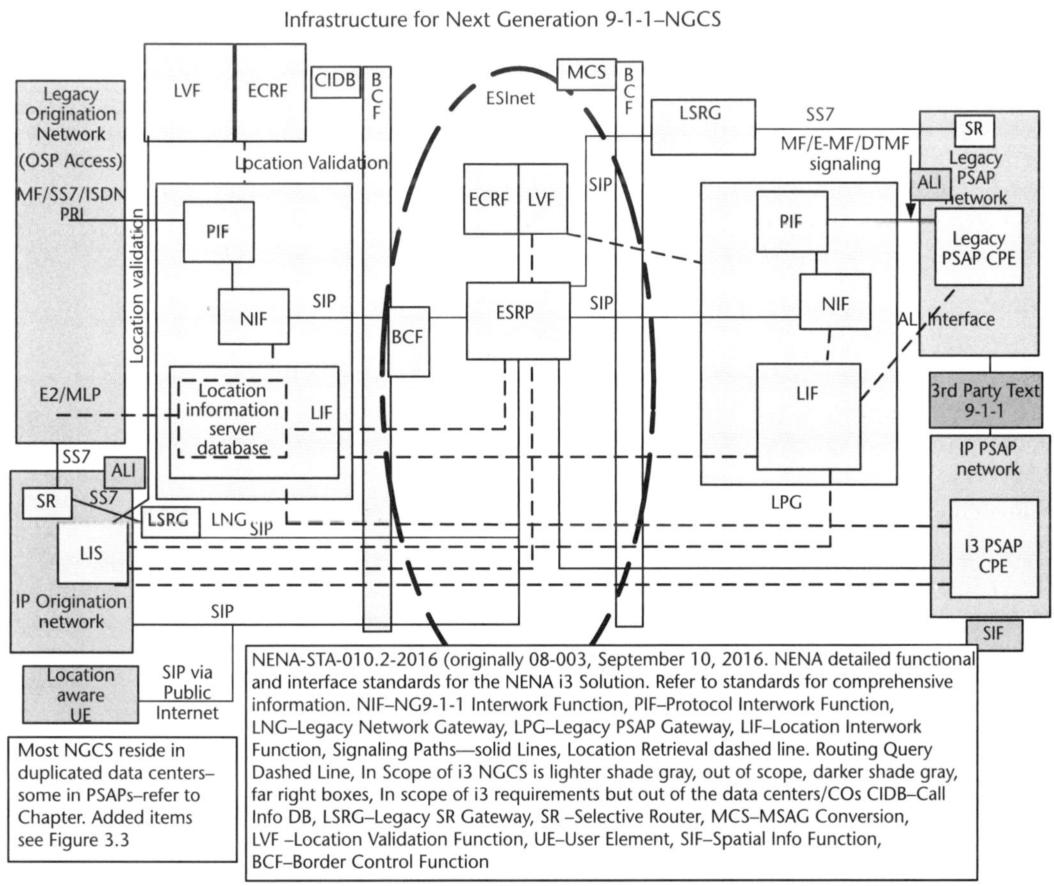

Figure 3.2 Data center example.

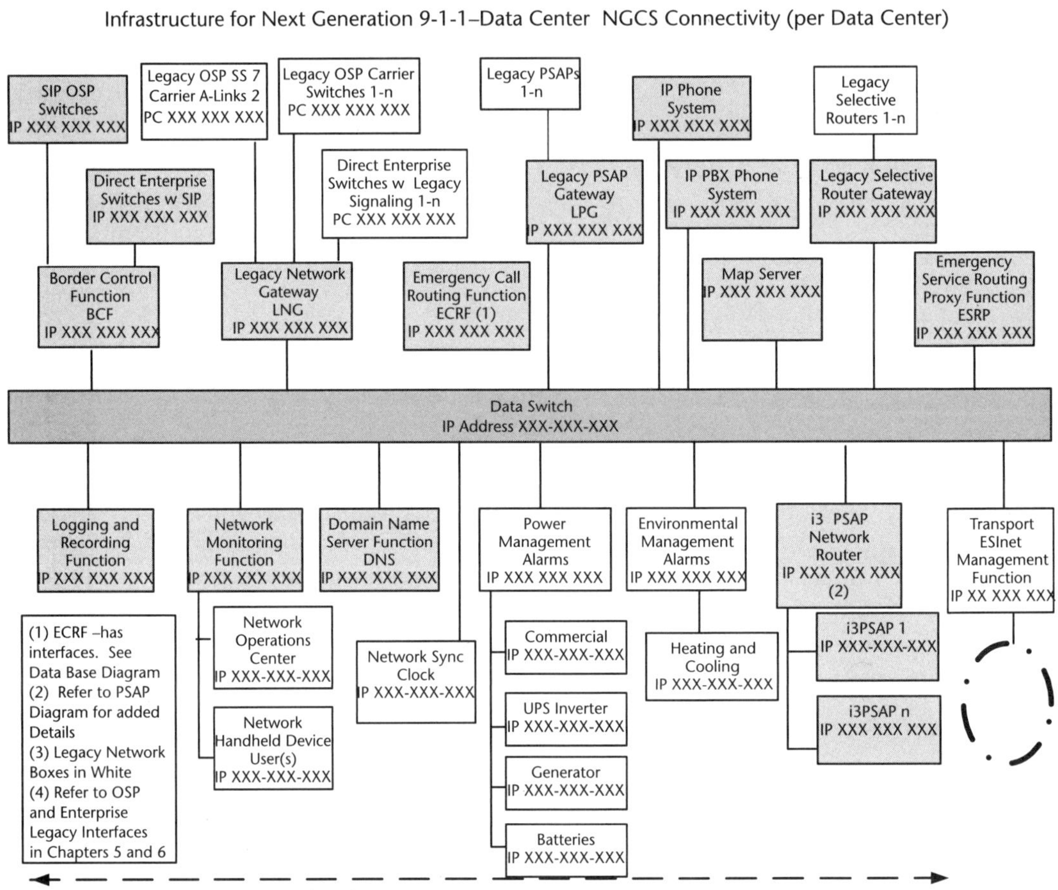

Figure 3.3 The NGCS architecture.

3.4.1 Security

The entrance to the NGCS is the border control function (BCF), a smart firewall that provides logical security to and from and across the ESInet. The BCF marks calls with suspicion levels and has functions to block specific call sources to prevent distributed denial of service attacks (DDoSs). Certain BCFs can withstand the largest feasible attack, currently known to be in the range of 10 Gbps. As with any NGCS component, the size and power of the BCF is proportionate to the geographic scope of the project.

BCFs, interchangeably called session border controllers (SBCs), provide the access control for protection from unwarranted cyberattacks, DDoSs, and/or focused network overloads. The design is in accordance with the NENA i3 standards; the data centers are the destination points for all inbound emergency NG9-1-1 traffic in the areas served. Further, the data centers are the egress points for all traffic outbound from the ESInet, such as outbound calls to and from one of the PSAPs serving the area to an adjacent PSAP, not yet on an ESInet. An emerging role is to protect one ESInet from another as ESInet connectivity is enabled.

3.4.1.1 Location Validation and Routing

One of the major advantages of NG9-1-1 is routing the call to the right PSAP based on the location of the caller. This is one of the ways to achieve the "any device, anywhere" goal of NG9-1-1. Routing the call involves several steps, which are laid out in the NENA requirements, and many of these steps pass through designated FEs. In the real world, many of these steps or functions are not, in fact, using separate pieces of hardware, and sometimes they are not using separate software.

The NG9-1-1 system accommodates locations other than a traditional physical address, specifically a GIS location. The NG9-1-1 location validation and routing function must use the information it receives about the caller's geographic location to route the call. As 9-1-1 evolved to E9-1-1, the ability to route the call to the PSAP and deliver the caller's address location information were done with tabular databases. In E9-1-1 terminology, the master street address guide (MSAG) is a database of street addresses and specific physical address ranges. The MSAG is used to validate customer address records in the ALI database. If the records do not conform to the address range, community, state and system edits, they are rejected and sent to a database manager and/or GIS administrator for resolution. If the records conform to the MSAG standards, the records are added to the ALI database. In NG9-1-1, these functions are replaced with several functions with new names and abbreviations that accomplish the same basic tasks as E9-1-1 but with more flexibility and interoperability. With NG9-1-1, GIS becomes the primary means of addressing and routing, replacing the historical legacy tabular records. The various vendors offering the NG9-1-1 FEs deploy their systems in different forms in terms of software running on separate servers, or they may in fact require separate software modules for each NENA-defined function running on common hardware. Some vendors may bundle the functions into related modules or as standalone programs on standalone servers.

Ultimately, both E9-1-1 and NG9-1-1 systems need to perform address validation and call routing based on the address or location of the originating caller. At a very high level and in a nutshell that is what the emergency call routing function (ECRF) does in conjunction with the emergency services routing proxy (ESRP). Once the ECRF receives the call information, technically from the BCF, it validates and determines the address/location, and the ESRP routes the call to the correct PSAP.

3.4.1.2 ESRP

The legacy SR function is handled by the ESRP, which is often referred to as the IP SR; some standards experts do not approve of that term, although it aids in understanding the function. The ESRP is the heart of the NG9-1-1 routing used for all calls.

3.4.1.3 ECRF

The ECRF provides the instructions for call routing. GIS data is provisioned to the ECRF. The ECRF theoretically can be queried by both the ESRP and PSAP workstations. The ECRF is queried using a protocol called location-to-server translation

(LoST), and the ECRF is considered a LoST server in IP networking terms. The LoST protocol is used in the IP and specifically in SIP calls. After validating with the location validation function (LVF), the ECRF sends either civic address or geo-coordinate location information in a data object known as the presence information data format (PIDF). The ECRF provides a location to route the call in the form of a uniform resource identifier (URI) and a service uniform resource name (URN).

The LVF is defined as a LoST protocol server where civic location information is validated against the authoritative GIS database information. It replaces the function of the MSAG. A civic address is considered valid if it can be located within the database uniquely and is suitable to provide an accurate route for an emergency call. The LVF may reside in a standalone server or be bundled with the ECRF.

The policy routing function (PRF) is a set of rules, usually defined by the 9-1-1 authority that deals with how the ESRP routes calls. These rules include but are not limited to the PSAP system state, congestion conditions, security posture, call suspicion, and call state. For example, rules are in the form of IF "this is true," THEN do "that." "This" represents the input conditions expressed with "and/or" statements. "That" is the route, the actual PSAP, the diversion PSAP, or busy. The PRF is dynamic, which means that the superuser has the capability to change it at any time and to have those changes applied immediately.

The N9-1-1 system has capabilities and functions for call handling that are not possible in E9-1-1 technology. Routing policy rules can intelligently route calls depending on conditions agreed upon with member PSAPs. This can range from a simple overflow of calls where all the workstations in each PSAP or county are busy handling NG9-1-1 calls. The additional calls can overflow to designated PSAP(s) until workstations become available to the long-term abandonment or isolation of a PSAP, and call routing would be long-term.

3.4.1.4 Management Information Systems (MISs)

Activity that occurs in the NG9-1-1 system is captured in the MIS portion of the NCGS. This core function of any NG9-1-1 system produces real-time data to enable the call/text logging, reporting, and troubleshooting. The MIS, typically an application that resides in the ESRP, contains information about the call, such as the start of the call, the answer time, and the call release. Various vendors' applications differ slightly in the user interface, the amount of information capture, and the ability to customize the data display, but all provide essential data to reconstruct the call events and enable the call/text to be recorded for future use. The storage mediums, length of storage, and number of backups are configurable and depend on state/local statutes or regulatory requirements.

3.4.1.5 Logging and Recording Functions

Recording NG911 voice calls is a standard practice. The recordings of 9-1-1 calls are used in criminal investigations and court cases and for distribution to the public. In the N G9-1-1 environment, recording text, images, and video is required. One major issue besides the security of the recordings is the storage of this information. In M ISs, the event records are of a manageable size, and the physical space to store the data is no longer a concern given the cost of digital storage. When video and images are added to the NG9-1-1 system, the requirements for storage and retention can become a major issue in terms of cost and manageable access to the data.

3.4.1.6 Default Routing

An agreement for default routing is another part of the NG9-1-1 decision-making that is identified in the appropriate IGAs. This agreement theoretically could extend to default routing, routing where insufficient or no location information arrives with the calls and where, typically, the call just has the ANI. Although receiving only the ANI is a rare event, it can and does occasionally happen. An example of default routing where only the ANI is received is where the SO wireline update has not arrived for the database and service has been activated. This situation can be mitigated by building NPA/NXX[15] tables to "more intelligently" route these calls with only an ANI to the best choice of PSAP; the wireline example should be used for only a matter of hours. Existing IGAs are documented between the public safety agency members to utilize the flexibility of the new system in routing overflow calls. Chapter 11 addresses IGAs.

To summarize how these FEs work, when a call comes from an OSP as described, the call goes to the ESRP via the SBC. The ESRP queries the prepopulated location information server (LIS) database and, if needed, goes to the mobile positioning center (MPC), the gateway mobile location center (GMLC) system, or the VPC to rebid and see if there is better location information. The ESRP queries the ECRF to determine which PSAP should handle the call. The ESRP gathers all relevant information sending the call with the data to the PSAP for answering. In the NG9-1-1 network, the PSAP equipment no longer is required to "dip" into or query the MPC or VPC ALI databases.

3.4.1.7 Gateways

LNG

Gateways are devices for the NGCS to interface with legacy technology. Gateway devices enable the NG9-1-1 technology to work with the legacy technology until it is retired.

OSPs using legacy call signaling are accommodated through at least two sets of legacy network gateways (LNGs), one in each data center. The LNGs convert SS7, ISDN PRI, and MF, as needed, to SIP before the calls are handed off to the NGCS. Refer to the call flow diagrams in Figures 5.1–5.4 and Figures 5.7 and 5.8, which reflect access to and egress from the NGCS. Readers can also refer to Figure 3.3, which addresses the NGCS as noted.

There are entry and exit points to and from the ESInet that exist as long as there are non-IP communications devices in the network. The entrance points from outside (inbound calls) comprise the legacy network gateway (LNG).

15. NPA: Refers to an area code in the North American numbering plan. Three digits in the form of NXX where N may be any number 2–9, and X is any number 0–9.; NXX: exchange code; -N is a number 2–9; X is a number 0–9. Exchange identification or central office code.

The LNG serves as the bridge between the existing originating networks and the BCF to gain access to the core of the NGCS. The LNG is on the access side of the network architecture. One often used interim means of interface to the LNG for transition purposes is from the existing legacy SR interface to the LNGs. This is an initial step to bring the CAMA/MF, SS7, and ISDN PRI (legacy signaling protocol) interfaces to the ESInet. The LNG is always outside the core NGCS. LNGs generally reside in data centers housing other redundant ESInet connections and NGCS FEs. The LNG routes calls via the ECRF, always through the BCFs.[16] Specifically, the LNG uses the ESRP to route the calls. The LNG interworks location protocols and formats between the legacy network and the NGCS. The e2 interfaces (wireless and VoIP) or internal LIS (which replaces ALI data for wireline) faces toward the legacy network. The LNG either supplies location-by-value in the SIP signaling or may supply a location reference that resolves to itself using SIP or the HTTP-enabled location delivery (HELD) protocol toward the ESInet. This is part of the design as long as legacy networks are deployed.

Legacy PSAP gateway (LPG)

The LPG is the entrance point for a PSAP using older workstations. LPGs allow existing non-upgraded PSAPs to connect to the ESInet and NGCS.

When public safety converts to NG9-1-1, and it is a purely IP environment, LPGs are no longer required. For the foreseeable future, public safety and the teams installing the NG9-1-1 systems must interface with non-IP/SIP calls. These calls originate using PSTN legacy telephone service, which uses a technology called time-division multiplexing (TDM). These non-SIP TDM calls must be converted to IP/SIP to enter or ingress the ESInet. Calls leaving the ESInet (egress) must be converted from IP/SIP to TDM. These gateways operate in both directions; ingress and egress and allow the new technology to interface with the legacy PSTN.

LNGs and LPGs are part of the architecture as long as legacy networks are deployed. Section 4.4 discusses communications and interfaces with neighboring public safety jurisdictions.

MAP

Mapping the location of a 9-1-1 caller has been a function of public safety since the era of hardcopy maps. Technology has enabled a map to be digitally displayed on a workstation, and in the case of E9-1-1, a caller's location is determined by the ANI received with the call and a look-up of the ALI database to retrieve the location. NG9-1-1 incorporates GIS data into the system for call routing, but it folds perfectly into a mapping system to locate the 9-1-1 caller's location. Multiple vendors

16. Figures 5.1–5.7 show each type of access and the interfaces suggested for NG9-1-1. Some implementors have located the LNGs within statewide NG9-1-1 data centers. In other implementations, the implementors decided to co-locate the LNGs at the same wire center hosting the current legacy SRs, because many OSPs have transport connections to those locations. This can be more expensive for the 9-1-1 SSP, and those OSPs seldom end up with redundant diverse OSP access to NG9-1-1 service. There are cost-benefit issues for each decision. In the Figures 5.1–5.7 drawn with this book, the LNGs reside in the data centers co-located with the BCFs and the NGCS. Without diverse access to NG9-1-1, is it truly NG9-1-1? Commissions will have to weigh in on these options and determine what is allowable, who will pay, and from what funds. Chapter 12 addresses finances.

offer mapping systems with differing functionality. Some public safety agencies use maps in the public domain such as Google Maps. Many PSAPs have outdated mapping systems that may be hard to integrate with NG9-1-1. Across the board, maps need to work for any PSAP to be able to back up any other PSAP on the ESInet.

Interworking functions

Interworking functions are the interactions between either the LNGs or the LPGs and the FEs of NG9-1-1. These functions convert or format the data for use by the NG9-1-1 FEs.

The spatial information function (SIF) is the overarching spatial function to interface NG9 1-1 with an existing GIS to provide location data to the LVF, ECRF, and ultimately, the map. The SIF, which replicates and provisions the data layers to the ECRF, is defined in NENA-STA-010.2-2016.

The protocol interworking function (PIF) is a component of an LNG or LPG that interworks legacy PSTN signaling such as ISDN-UP trunking or CAMA with SIP signaling.

The NG9-1-1 interworking function (NIF) component is used for interworking with the ESRP for routing instructions. This function is in either the LNG or LPG. The location interworking function (LIF) is the component of a LNG responsible for taking the appropriate information from the incoming signaling (i.e., calling number/ANI, ESRK, and cell site/sector) and using it to acquire location information that can be utilized to route the emergency call and to provide location information to the PSAP. In an LPG, this functional component takes the information from an ALI query and uses it to obtain location from a LIS. See Figure 3.2.

Legacy selective router gateway (LSRG)

The FE is a temporary solution to interface the legacy 9-1-1 SRs and E9-1-1 switch technology used by the telephone companies with the NG9-1-1 architecture. The LSRG allows calls to be routed or transferred between the TDM legacy SR and the IP-based NG9-1-1 networks. Once a public safety agency and its neighbors convert to the NG9-1-1 technology, the legacy SR(s) can be decommissioned, and then LSRG(s) can be removed. The challenge is split exchange routing. Section 5.11 details the LSRG.

Timing source

As a call progresses through the NG9-1-1 network, it is essential for all parts of the call to be time-stamped from start to finish using the network time protocol[17] (NTP). Each data center should have access to the network clock for synchronization interfacing the OSPs and the ESInet and the NGCS and other FEs inside the 9-1-1 SSP's virtual or physical space. Each PSAP should have access to the data center or telco wire center clocks for synchronization to ensure that time stamps are accurate.

17. NTP is a networking protocol for clock synchronization between computer systems over packet-switched, variable-latency data networks.

Switches and routers

NG9-1-1 is 100% IP-based. The architecture has a variety of data switches and data routers to move the data to the right location based on the function being performed and the IP addresses of the FEs. Data switches are also assigned IP addresses, and they are the devices using the datalink layer, layer 2, of the OSI model. Typically, data switches are a multiport network bridge that forwards data to the devices that need it. The data switches can also act as a port multiplier, like an electrical outlet strip.

Data routers are located in data centers, and at PSAPs, and they operate on layer 3, the network layer, of the OSI model. Routers allow dissimilar networks to essentially serve a routing bridge even for a subnet. There are smart switch and router combinations that can perform the same function as separate data switches and data routers, which typically have more configuration options.

OSP direct access

Direct OSP carriers connect to the data centers using a POI as described in Section 3.2 ESInet. The POIs can be located in the data centers or at strategic points on the ESInet. The POI is a location at which the OSP can terminate circuits for its 9-1-1 traffic and any equipment associated with the circuit. Each OSP chooses from legacy (SS7/ISDN-PRI) and newer protocols for trunk signaling (SIP). Where the ESInet appears in an OSP or private-switched ALI (PS/ALI) switch location, the OSP or PS/ALI customer may connect to the fiber at that location, thus including a zero-mile access solution. Since there are fiber-optic providers that already have many large customers on their rings, this an ideal option. Chapter 6 discusses enterprise-direct customers.

For access, the 9-1-1 SSP cannot measure blockage from the originating OSP's perspective. Cooperative efforts are needed to size the individual paths from each OSP after approval of the plans at a high level. The incoming devices are either the LNGs in a legacy environment or the SBC in a SIP access environment. The 9-1-1 SSP needs to size the NGCS accordingly to ensure sufficient capacity. Since not many NG9-1-1 networks are utilizing direct traffic, the main interfaces are from legacy SRs into the LNGs. In those cases, we recommend that CAMA trunks be replaced by ISDN-PRI trunking. In other cases, third-party aggregators have stepped up to convert the legacy SR signaling to SIP before sending it forward to the 9-1-1 SSP. In many cases, although approximately 100 OSPs may operate in an ESInet footprint, the number of actual connections can be much lower when aggregators are involved. Testing is often managed by the aggregators on behalf of their clients. It is highly recommended that each OSP be identified and tested even when pathways carry traffic aggregated from many OSPs. End-to-end, differences in the OSP originating technology, depending on age and manufacturer, can result in call failures.

When new NG9-1-1 networks are tested and turned up, we suggest that traffic load be added gradually and monitored for performance until experience can be gained to avoid blockage anywhere along the network. Chapters 7 and 8 discuss cutover.

There is a lot of other traffic (calls and data) and transfers passing back and forth along the ESInet. The ESInet needs enough capacity to send calls and datalink queries for location to PSAPs and accept them from PSAPs.

User equipment (UE)

The term UE is used to describe a wide range of devices by NENA and other industry players. NENA defines UE as “a device allowing a user access to network services.” It further defines UE in NENA’s Functional and Interface Standards for Next Generation 9-1-1 Version 1.0 (i3) NENA 08-002, Version 1.0, December 18, 2007 as “the 3GPP[18] term for IP client or endpoint.” In the same document, describing the IP client, NENA says, “This term is used to refer to the IP endpoint communications equipment or application that is used to originate a voice, video or text request for emergency services (e.g., by calling 9-1-1). The term IP device or IP endpoint may also be used. Also, in the same document,” NENA writes that, “IP clients may request routing information from ECRFs, or a proxy (e.g., a VoIP service provider) may request this information on their behalf.” NENA provides diagrams in the document referenced above as “Wireless/IP Client” (NENA Master Glossary of 9-1-1 Terminology NENA-ADM-000.22-2018).

3.5 PSAP

A PSAP’s primary function should be to provide citizens with high-quality 9-1-1 services. PSAPs are connected to the ESInet, NGCS, and FEs, which are designed to provide PSAPs with the information they need to achieve that objective. Figure 3.4 shows how PSAPs are connected to the ESInet and each other and the equipment that remains.

3.5.1 Workstations

Workstations in a i3PSAP are off-the-shelf computers with software installed to connect to the ESInet to interact with the BCF, NGCS FEs, location databases, other PSAPs, and other ESInets. Increasingly the computers have multiple video cards to support multiple monitors, so the call taker screen is on one monitor and the map or CAD on another. This increases the visibility of the necessary elements to support call answering and rapid dispatch. Workstations configured on computers allow customization possibilities not available in E9-1-1. Sounds can alert the call taker to incoming calls and the layout of the screens can be modified locally.

3.5.2 i3 PSAP

PSAPs need to meet certain requirements to be a fully functional and operating i3 PSAP. Most PSAPs operate 24 by 7 by 365. There are exceptions, but usually that includes small PSAPs in rural locations. Practical matters, such as expense, may eliminate these operations as the trend is to consolidate. PSAPs adopt a set of

18. 3rd Generation Partnership Project; see Glossary of Terms and Acronyms.

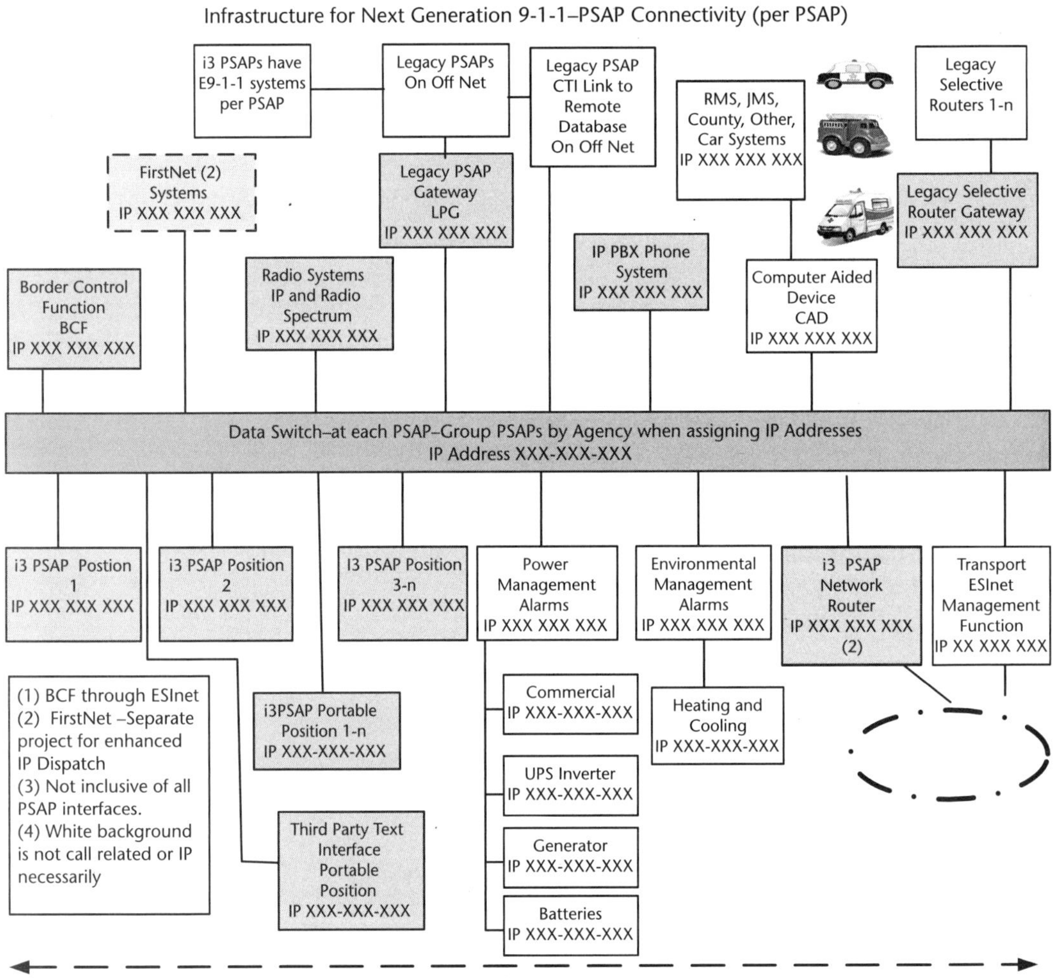

Figure 3.4 PSAP connections and equipment.

standard operating guidelines that include PSAP privacy and security. The BCF can be virtually extended to keep a PSAP logically secure.

PSAPs back each other up. NG9-1-1 protocols allow more rules for sending calls to the backup PSAP. All PSAPs on the ESInet can operate as one large PSAP with rules governing the backup protocols. PSAPs can be staffed for the busiest hours only or the busiest days. IGAs are a necessity when load sharing between PSAPs. Figure 3.4 shows a multiplicity of PSAP interfaces; note that the depiction is an example that does not represent any PSAP in particular.

3.5.3 PSAP Timing Source

Each PSAP component should have full access to the data center clocks for synchronization to ensure time stamps are accurate.

3.5.4 Backup Electrical Power

A PSAP requires an alternate power source, or generator, that is capable of sustaining full operation of the PSAP, including communications, in the event of a commercial power outage. Each PSAP should have an uninterruptible power system (UPS) to allow continued operation during switch over from commercial power to alternate power. PSAPS should maintain detailed power and grounding surveys. Each PSAP should be verified to have a properly engineered building ground in place, appropriate heating and cooling, physical and logical security, and fire-suppression equipment.

3.5.5 MIS Call Recording

A call-logging recorder must be accessible to be sure that the PSAP is capable of recording and date- and time-stamping all 9-1-1 calls per position, texts, and radio frequencies. State and local laws usually require the PSAPs to maintain an electronic log of the 9-1-1 calls and texts. In the IP-based NG9-1-1 environment, all events including routing and transfer information are recorded in the embedded MIS. Generally, that system is duplicated and resides in the data centers. In some interim situations, PSAPs and public safety agencies are choosing to retain their own local logging and recording systems. The 9-1-1 SSP and its chosen SI need to work closely with the PSAP managers to be sure that all functions operate as required by law in their own jurisdictions.

3.5.6 Text to 9-1-1

Cellular or wireless phones have become the predominate method of calling 9-1-1, and that trend continues to gain momentum as wireline service declines and wireless technology improves. Cellular phone users increasingly expect to be able to send a text to 9-1-1. The PSAPs decide when they are technically ready to receive texts; PSAP training and public education is required before implementation. The PSAPs must notify the FCC and in turn, ensure that the wireless OSPs set up dates to test and complete the provisioning process. There are three types of interim text-to-9-1-1 solutions. The IP solution completes calls by connecting to the NGCS in the data centers. Local solutions connect to each and every PSAP one at a time. This is an example of a nonstandard standard. Chapter 10 addresses text to 9-1-1. Since texts are as important as any other emergency calls, their progress and exceptions need to be monitored and the 9-1-1 SSP and the PSAPs alerted if texts are failing at any point along the network.

3.5.7 TTY/TDD

TTY is a legacy technology from the 1960s. In the 9-1-1/E9-1-1 world, TTY is a standalone device that the hearing-impaired use to reach emergency services. In a NG9-1-1 environment, TTY calls come through the NGCS and are delivered to the PSAPs on the call taker's workstation like a voice call; they *do not* require a separate device. Texting from cellular phones to 9-1-1 is a replacement for TTY. There is a push nationally and internationally to implement text to 9-1-1, but there is no single solution.

3.5.8 Disaster Recovery and Incident Command

In case of a PSAP needing to abandon the physical space it typically occupies, the PSAP can stay operational in the NG9-1-1 environment with the use of portable workstations. Portable workstations are heavy-duty laptops in a hard case that can be transported anywhere. The portable workstations can be used as command post workstations or deployed in various situations. Since the system is IP-based, these fully functional portable workstations can connect to the ESInet and receive calls.

3.6 Dispatch

Functionally, call taking, answering 9-1-1 calls, and dispatching resources to respond to the caller may be functions performed by either the same individual or multiple people. This addresses the function of dispatch, whether standalone, integrated, or remote.

3.6.1 CAD

CAD is proprietary software to track available resources, dispatch those resources, and maintain situational awareness of resources operating in the field. To work effectively in the NG9-1-1 environment the CAD must be IP-enabled. E9-1-1 PSAPs already have access to CAD systems. A CAD system takes the feeds for the various FEs and systems attached to output ports. Some NG9-1-1 vendors have a serial port feed for non-IP enabled CAD systems. CADs are usually associated with a suite of other products such as records management systems (RMSs) and jail management systems (JMSs). CAD systems hold a significant amount of historical data that is vitally important to public safety. Being proprietary software, moving from one CAD system vendor to another can be a significant event in terms of losing access to the data or paying to have the data converted. The CAD vendor losing the business typically charges significantly to make the data available to the new system. CAD systems increasingly connect to field personnel utilizing mobile data terminals (MDTs). CAD is not considered a part of the suite of NG9-1-1 NGCS or i3PSAP products, but it is essential for the new i3PSAPs to be fully integrated with whatever CAD is identified for 9-1-1 SSP system integration in the RFP.

3.6.2 Radio and Telecommunications

All emergency services usually have the capability to be linked either by radio or telephone. Details on both primary and secondary communication procedures are contained in the call-handling IGAs. Most regulatory commissions require a document outlining participating and adjacent agencies for each PSAP; this document must include the radio and other dispatch capabilities. To be approved in some jurisdictions, the 9-1-1 SSP must ensure all participating agencies and adjacent agencies have their telephone numbers listed for emergency seven- or 10-digit calling using a single button and identified by radio call sign for dispatch. This may be called additional data in NG9-1-1 parlance, and it is required to be loaded and tested before a system can turn up for service. In many cases transfer trunking, detailed in Chapter 4, may be required.

Details of the dispatch system utilized by the PSAPs are included in the call-handling agreements discussed in Chapter 11 and in the examples of participating agency and adjacent agency and communities served documents in Chapter 4.

3.7 Database Management and GIS

The database management software, which should fully integrate with the other NGCS of the NG9-1-1 system, performs data validation, request/response functions, and data-quality management and provides real-time communication tools to aid in data management.

DBMSs should correlate postal, MSAG, and GIS data. This matching process can primarily be done with vendor software, but some of this correlation may require manual intervention. The more highly correlated the data is the more accurate call routing will be.

GIS coding should be added to all OSP SO records that are used to update the location ALI databases, so that the NG9-1-1 system can accurately route the call based on location.

Every structure should be identified, and every residence assigned a street number with a street or road name. A geospatial value must be assigned with address points to identify the residence/business where the 9-1-1 request is made. All database elements should comply with NENA data formats for ALI, MSAG, and GIS.

9-1-1 SSPs are responsible for all functional aspects of database operations and need to coordinate with all OSPs to receive updates and maintain the database(s). They also need to ensure that the database records received from the OSPs are reviewed for location accuracy and validated and integrated into the active working records. Most wireline customer records come from the SO process of the OSPs. The OSPs transmits the database updates to the 9-1-1 SSP, which works closely with each OSP directly on database updates.

Split-exchange call-handling records are handled in the same manner. SO updates to the LVF (ALI) records are accomplished using database management software to process updates from all OSPs.

Wireless and mobile VoIP callers are located via third-parties, and the location of the caller is determined by these third-party MPCs and VPCs with the aid of the OSP. Cellular and mobile VoIP calls use pANI records to denote which MPC or VPC is needed to determine location.

3.7.1 Converting Existing Database/MSAG Records Versus Building from the Ground Up

There has been a debate over who "owns" the MSAG and ALI. The MSAG defines the address ranges for each route (e.g., street, road, and highway). It contains data elements such as community and state to add uniqueness to the records that aid in the ECRF routing decision. Addressing authority is usually a county or municipality function. This makes the MSAG the logical responsibility of the public safety agencies. There are cases where public safety has handed over this function to one or more traditional landline telco/9-1-1 OSP. In a NG9-1-1 environment, particular attention must be paid to the MSAG; in fact, because of its critical role in building the ECRF and the subsequent routing decisions, the MSAG should be owned by the public safety agency. The debate of ALI database ownership is as old as 9-1-1. Depending on state law, regulatory rules and tariffs tend to favor the 9-1-1 SSP as the owner since the owner can charge significant amounts of money to provide copies or images of the ALI database. Public safety argues that it has paid for the OSP to accumulate and maintain the ALI database. As a practical matter, the 9-1-1 SSP controls access to the ALI database, and, unless specified by law or regulatory rule, there is a charge to obtain a copy of the ALI database.

When converting to NG9-1-1 there are choices between converting the existing records or building the necessary data elements from scratch. Key elements in the decision would be the source of the existing ALI database, the accuracy of the data, and, possibly, the cost. In many cases the existing 9-1-1 SSP is the NG9-1-1 SSP, and converting the existing data is probably the obvious choice. Although converting existing data has the cheapest upfront cost, it is important also to consider the quality and accuracy of the data. Subpar data will result in routing problems, prove to be more expensive in the long run, and expose the public safety agency to liability. It's not unheard of for any OSP's service representative to "force" an address into the database by either changing the MSAG to allow the order to "post-complete" or using a postal address that may or may not conform to the new NG9-1-1 standards in terms of data elements like community or address prefix and suffix abbreviations. This mismatch of the SO data, the MSAG, and GIS data will not allow the ECRF to resolve the destination of the call and thus result in a default route that will likely need to be transfered, costing time, which can cost lives or property.

An alternative is to create a new LVF/LIS (ALI) database by collecting records from all wireline and wireless MPCs, VPCs, and VoIP providers that offer a service that allows customers to reach 9-1-1. OSPs would be expected to provide a fresh/scrubbed load of NENA-compliant location records for each of their customers. Many or, in fact, most states require OSPs to provide records and updates to the public safety agency or their surrogate the 9-1-1 SSP.

The same options exist for the MSAG: convert the existing MSAG or build the LVF from scratch. The difference with the MSAG and the OSP location data is that the public safety agency is responsible for the MSAG or newly named LVF.

3.7.2 Steps to Test the Database Prior to NG9-1-1 Cutover

The 9-1-1 SSP obtains the ALI image copy from each legacy 9-1-1 OSP and/or directly connected PS/ALI customer two weeks prior to live traffic being delivered to the NG9-1-1 system. This image copy is taken after the workweek's updates have been processed to make the database as current as possible.

Updates are sent daily, each business workday, from each OSP directly to the 9-1-1 SSP, to update the database with the information sent from the OSP.

Records that fail the database edits for syntax or formatting errors are returned to the submitting OSP for correction. Error records sent to OSP and PS/ALI customers must be tracked for resolution and completion.

After 10 days of running concurrent updates of the legacy 9-1-1 OSPs and PS/ALI customers' databases with the identical updates processed into the NG9-1-1 database, the legacy 9-1-1 SSP provides a current, updated, E9-1-1 version of the ALI database that is compared against the NG9-1-1 database that has been updated in parallel.

Discrepancies must be researched and resolved, depending on the situation, with the legacy 9-1-1 SSP or the individual OSP's enterprise PS/ALI customer that controls the records with the discrepancies.

Test results are determined by reviewing the differences in the database on a record-by-record basis, based on update activity.

When these tests are successful, 9-1-1 SSP will continue to process the NG9-1-1 database updates sent from the 9-1-1 OSPs until live traffic is successfully cut, beginning within days of successful database testing. Chapter 7 discusses overall testing, while Chapter 8 addresses cutover processes and options.

Once live traffic is cut, the 9-1-1 service provider processes the respective OSPs' and PS/ALI customers' updates daily.

3.7.3 LIS/LVF Updates

The SO updates that are managed by the 9-1-1 SSP daily should utilize multiple logic checks that guard against invalid ALI updates and increase the reliability of data. These invalid updates would include, but are not limited to, ensuring that the address conforms to public records, including GIS ECRF data (which replaces the MSAG), and that the ALI record telephone number is in a valid format. The database system should support direct SO-based data entry, manual updates, as well as the import and export of records that conform to the NENA format. Updates are usually sent via a file transfer protocol (FTP) site. ALI delivery/updates and corrections should be accomplished via electronic transmittal although some smaller OSPs or regional incumbent local exchange carriers (ILECs) still use paper or fax and require manual updates. The 9-1-1 SSP should accept all updates from the OSPs and, after review by the system, should update and maintain the NG9-1-1 databases. The public safety managers and their GIS coordinators are responsible for maintaining the geographical boundaries, the format of the records, and additions or changes to street addresses or other major features of the physical geography over time. Housing additions and road additions for example are updated by the public safety managers into the database management system.

One of the longest lead times in planning and implementing NG9-1-1 is the preparation and validation of the NG9-1-1 databases and GIS records. Negotiations across the parties involved in the new geocentric NG9-1-1 system may be challenging and need 9-1-1 SSP and vendor leadership and support to meet the timelines and goals of the project and to stay on budget. Involved parties need to consider a set of standard naming conventions, but the practical issue is counties have their own naming conventions they are reluctant to change. ESInet to ESInet adds another level of complexity. This gets back to the premise of standards by committee that are not enforceable. To start with, a whole state needs to work seamlessly. At state boundaries, things need to interoperate seamlessly and with high security. Country-to-country interfaces are another layer to be negotiated and tested carefully before interfacing outside country boundaries.

3.8 Power, Grounding, and Environmental Factors

Whether it is NG9-1-1 or any communications network, there are certain vulnerabilities that must be considered before implementing an important service. The environment in which you choose to install and maintain the components of any network must be secure and reliable and address environmental concerns. Chief among factors to consider are power, grounding, and timing or synchronization and overall environmental issues including heating, cooling, and fire suppression.

3.8.1 Power

The most critical item for the NG9-1-1 infrastructure is power. As you start your journey, be sure to verify at each point along the way that adequate power is in place. It is advisable to have an inventory handy for planning implementation, ongoing operations, and maintenance. It is crucial that the power that feeds the equipment is on a separate circuit and breaker from the power used by maintenance people or staff in the data center/PSAP for appliances like toasters and microwaves. It is important to keep a clean duplex power supply to the equipment without interference from other electrical devices.

There are four main categories required: commercial power, uninterruptible power, generator power, and battery power. The 9-1-1 SSP and each data center and PSAP manager need to have an inventory and phone numbers to reach for power support during an emergency. A routine maintenance log needs to be filed with the 9-1-1 SSP monthly for each type of test and maintenance, including refilling fuel or adding water to batteries. Table 3.1 lists examples of maintenance items. All power sources that can be alarmed need to have alarms extended to the 9-1-1 SSP NOC even if in data centers and PSAPs managed by others. The power element information, testing, and maintenance schedules need to be added to SLAs in contracts.

Table 3.1 Monthly Maintenance Log

Location: Data Center, PSAP, or ESInet Hub Location			
Commercial Power	Company Name	Location in Building	Contact Number
Uninterruptible Power	Name of Unit/Size	Location in Building	Contact Number
Generator/Fuel	Name/Size /Fuel Source	Location in Building	Contact Number(s)
Batteries	Name/Source	Location in Building	Contact Number
Fire Suppression	Name/Source	Location in Building	Contact Number
Power Person/Site	Contact Number	Alternate/Site	Contact Number

3.8.2 Grounding

Grounding is another critical item that needs to occur *before* installing the first piece of NG9-1-1 equipment for all locations end-to-end. Grounding protects electrical equipment from damage caused by electrical surges or other types of faults. Grounding has two components. The first is the overall building ground. This requires a professional building ground engineer to inspect each site and certify that the buildings that house any part of the equipment for NG9-1-1 are properly grounded. Electrical storms often cause damage to buildings that travels through to equipment and computers. The resultant damage if grounding is not properly completed is loss of any and all cable, monitors, computers, systems, servers, and routers. The agencies that hire the 9-1-1 SSP need to determine whose responsibility it is to get this done. Usually the building owner ought to be accountable for grounding. The 9-1-1 SSP needs completed statements signed by the agencies for every location before work begins. Again, the SLAs in contracts need specifications to include grounding.

The next grounding component is for the floor, the racks, and the equipment inside each location. All this needs to be tied to the building ground. The handling of equipment is a sensitive matter, and no one should be touching the equipment without a proper wrist-style grounding strap. During ongoing operations, the PSAP call taker or data center technician might be asked to reboot a router or switch. The grounding strap should be attached to the grounded rack and be put on before performing the activity. If a new circuit pack or router is required, then the technician doing the work needs to perform it safely. The floors in equipment areas should be treated with electrostatic discharge (ESD) cleaning products. This is especially true in each data center. Cabinets ought to be mounted and equipment stored in a way, so they are not bumped into by any personnel

3.8.3 Timing and Synchronization

It is also essential for all the electronics end-to-end to be time-synchronized; otherwise, delays, jitter, and static can prevent the caller and the called party from communicating. Possible disconnects or failure to even get a call set up can occur if timing is not properly synchronized. The clock the United States uses for Tier 1 synchronization is in Denver, Colorado. All United States-based telecommunications companies trace their clock sync back to this source eventually. There are multiple tiers of timing extended through transport and switching, and ultimately NG9-1-1 needs to be in sync. The 9-1-1 SSP needs to provide for a central timing source consistent with the clock in Denver. The choice of the clock source is part of the RFP response. Each data center may already have a standard proper clock source that it can extend under contract, and the SLAs need to specify that clock source and maintenance are included, along with alerts and alarms, if the clock source fails. If the clock source fails, the "drift" starts almost immediately and can take down whole networks. Besides impacting service across the ESInet, "drift" can affect the records police need for legal matters that are used in trials and for testimony. Moreover, it certainly can impact the MIS information that the PSAPs use to make staffing decisions and reports to the commissions about holding times on calls and calls lost or blocked.

3.8.4 Environmental

3.8.4.1 Locations

Overall location considerations are essential: Do not choose locations for data centers or PSAPs in locations prone to problems, such as flood plains, earthquake zones, hurricanes, and wildfires, nor in any natural disaster-prone area or manmade terrorist target area. While it may be impossible to assess each possibility, take the utmost care to choose locations wisely.

At a minimum, choose data centers with enough distance between them to prevent a local disaster of any kind from being likely to impact the mirror-image data center.

Where possible, choose locations on a fiber ring for data centers; those with dual-wire entrances are ideal. PSAPs are seldom directly on a fiber ring unless there is a municipality with a private fiber network serving the public safety building, which is ideal. PSAP size and IGAs with overflow or alternate PSAPs constitute primary considerations where diversity is simply not available.

3.8.4.2 Physical Security

Physical security is mandatory at each data center and PSAP location. Security guidelines for each building need to be validated. Work personnel need to have to gain authorized access to the buildings 24 by 7 by 365. Yet, every person entering the building needs to be screened. A system should be in place to log entries and passwords, or passcodes need to be changed routinely. It is essential to know each person specifically who enters—and not to allow just anyone with a generic door or passcode to enter and leave. Alerts and logs need to go to the 9-1-1 SSP NOC and, ideally, to the managers of each building's 9-1-1 responsibilities in near real time.

3.8.4.3 Cybersecurity

In any data center or PSAP, the biggest threat to service is insufficient cybersecurity. NENA has been expanding its guidelines to comply with federal, state, and municipal guidelines. The BCF function is the "big ticket" item critical to the security of the network end-to-end. As with any tool, buying a firewall and installing it is not enough. The 9-1-1 SSP needs to have trained and qualified cybersecurity expertise on staff to work with the PSAP managers and agencies to implement thresholds for allowing traffic to flow through at a reasonable rate while blocking or diverting heavy loads or DDoS. The BCF is a tool that requires the expertise of a cybersecurity specialist to install it and ensure that it is protecting the whole environment. We have seen PSAP call blockage occur when a bored PSAP call taker decided to view a streaming video, causing blockage of the whole PSAP because the streaming used all available bandwidth. Cybersecurity means monitoring for spikes and setting up active real-time alerts. It is critical to test for load and overload before a network goes online. There are programs to create load, validate that the firewalls are not set too high or low, and verify that the alerts are actionable and sent to knowledgeable personnel. University and telecommunications laboratories have been helpful in testing the NG9-1-1 network and creating dynamic stress and reports before a network goes live (Chapter 10).

3.8.4.4 Heating, Cooling, and Fire Suppression

Heating, ventilation, and air conditioning (HVAC) are major considerations since electronic devices emit heat and require low humidity. Modern computers are not quite as stringent in terms of operations as their prior telephone company switching equipment, but inside each building, a physical check should be made of the HVAC, at the locations that house the equipment essential to 9-1-1. Each FE manufacturer has operating limits on the temperature and humidity at which its equipment performs optimally. Small rural PSAPs often have the telco closet spaces serving double duty as broom closets with no HVAC. Conditions need to be tolerable before installing equipment racks for the switches, routers, and other NG9-1-1 FEs. If the building does not have a temperature gauge inside the rooms housing the PSAP equipment, there are small portable battery-powered devices that can transmit using the internet of things (IoT) to the monitoring systems that should be available to the 9-1-1 NOC and the PSAP. Excessive temperatures can cause fires. Commercial power can go out, impacting heating and cooling even if backup power is keeping the PSAP online.

Data centers need to share their environmental support infrastructure heating and cooling plans to ensure that they are monitoring their temperatures and that they have backup heating and cooling options should there be a commercial power outage affecting the data centers.

Fire-suppression equipment must be identified in all ESInet and data center and PSAP locations. The types of equipment available needs to match the impact on the electronics, and the fire department rules for getting access to the building must be followed with floors and equipment marked accordingly. State guidelines often reflect rules.

3.9 Engineering and Capacity Considerations

Every NG9-1-1 network needs a lead qualified design engineer. That engineer may need a team of engineers depending on the topic, transport switching, translations, TDM and IT, signaling old and new, and IP engineers (i.e., transport).

Chapter 2 addresses skills needed by public safety agencies, the 9-1-1 SSP, the system, and the infrastructure. This section highlights the considerations for the size of the ESInet and paths from OSP and telco locations to the data centers and from data centers to the PSAPs.

3.9.1 ESInet

Recall the recommendation that the size of an ESInet should be no less than 100 Mbps and could be as large as 100 Gbps. So, how large does it need to be? Let's look at the demand.

3.9.1.1 OSP and Equipment to ESInet

The aggregate demand from all OSPs, the equipment and the PSAPs and any operations, administration, and management (OA&M), equipment needs to be considered. The old rule of thumb for OSP interfaces was that no less than two trunks in a trunk group from any switch is engineered to carry E9-1-1 calls to the legacy SR. Two trunks per 10,000 lines of customers in an originating switch was a good estimate until actual traffic could be measured. Telcos are not sharing their numbers of lines per switch with competitive 9-1-1 SSPs, and no commission is requiring them to do so. Accordingly, first the 9-1-1 SSP needs to trust that the OSP knows their own traffic loads to 9-1-1 today, and because the commissions usually require OSPs themselves to block no more than one in 100 calls in a busy hour of the busy season, the OSPs state the number of trunks required per switch, but no less than two. That two trunks to each legacy SR becomes two trunk groups to each data center if the OSP decides to connect diversely to the 9-1-1 SSP data centers. So far commissions have not been forthcoming on an OSP requirement to diversely access the NG9-1-1 networks. It is not an engineering problem; it is a business decision.

There are sometimes hundreds of OSPs in an ESInet footprint and many ways for the OSPs like wireless, CLEC, and VoIP OSPs to consolidate traffic into an area, saving them money. Some OSPs may need a T1 (or DS1) into each data center, and some even require a DS3 level of access. If OSPs have soft switches they might require anywhere from 10 Mbps to 20–100 MBPS of capacity to carry their demand. The Ethernet providers generally will not sell below 10-Mbps access pipes. The transport medium could be copper or fiber.

In much of rural America there are host and remote end offices with many remotes, and often those remotes have no dedicated 9-1-1 trunking and only one path to get to the host. The trunking at the host switch is incremented by the added callers calling from each remote they serve and the host's own callers. Remotes have almost no chance for 9-1-1 diversity, though a few have 9-1-1 bypass capability (although we have not seen it implemented). This is a simple fact of landline telco design: Facility diversity is severely limited in rural America.

The SS7 switches have A links connecting to their own STPs for all of a CO's access to anywhere—not just 9-1-1. We are not suggesting that F, or fully associated, SS7 links be added to reach the LNGs. The 9-1-1 SSP must be an SS7 STP provider or contract with an SS7 provider that may already be connected to STP pairs using D, or diagonal, links; then the new SS7 provider designs and engineers A links into each data center over diverse facilities. The SS7 provider needs the SS7 point codes (PCs) for all OSPs feeding traffic into the NG9-1-1 network. This is an added engineering requirement. Again, the existing OSPs have been reluctant to share the PC information from each OSP switch with the new 9-1-1 SSPs, but it must be obtained to connect the OSPs. The OSP engineers know this information. The regulatory commissions typically have not mandated this data exchange. Rather, it needs to be part of the commercial agreements between the 9-1-1 SSP and each OSP entering the network.

3.9.2 ESInet Capacity from all OSPs

So how do you size an ESInet when the data is hard to obtain and scattered across OSP engineering databases? The authors used the U.S. Census to create an initial estimate. By counting every person in the ESInet footprint and assuming that each may have a simultaneous need for 9-1-1 in the network busy hour and that each wants voice, video, and data, we can estimate a percentage that needs to pass through the ESInet to all PSAPs. That way, transient callers, concert, and venue groups and all can get access. We plan for each data center to be operating at 40% capacity, sharing the load, and if one fails the other data center takes over at 80% capacity. If there is a true crisis, then an overload could take a data center's NGCS to 100% and still not block any more than one in 100 calls. In the regional ESInets where we have experience, 100 Mbps gave us confidence and proved to be adequate. Data centers, components in the data center, and certainly PSAPs have failed. Never, however, has the NG9-1-1 system lost calls. Top-down ESInet engineering works. Actual data helps refine contractual arrangements. Accordingly, we recommend that connection agreements require annual or semiannual joint engineering reviews to validate capacity demand and establish a forecast for future consolidations, cutovers, and growth.

3.9.3 PSAPs

Facilities to the PSAPs can be measured based on the numbers and size of the PSAP workstations at each site. If you assume that each PSAP workstation position could handle voice, video, and data, that might consume up to 500 Kbps per call. That may be extreme. However, if you allow for that much demand per position, you need a T1's worth of bandwidth per PSAP for every three positions. Small PSAPs often fall into that range in rural America. Recall that a T1 is a minimum interface to the NGCS. If the trunk paths use SIP or Ethernet, 10 Mbps could handle it, and of course, that is usually the smallest amount of bandwidth to a PSAP that can be ordered. Use the number of positions times 500 Kbps to estimate capacity. The network gets more efficient as the number of positions grows. Engineering tables can be used to determine bandwidth, remembering that what can be purchased often jumps up at predesignated rates. Actual experience is the best guide. Remember that there should absolutely always be at least one PSAP that can handle the overflow and possible failure of the target PSAP. There is a default PSAP and backup default PSAP that could need added bandwidth if default routed calls increase because of the quality of the ALI location information. Some public safety agencies want to consolidate. There always needs to be capacity on the ESInet for one or more PSAPs to fail at the same time. Sadly, we have seen multiple simultaneous PSAP failures happen. Cable cuts, floods, and wildfires can cause abandonment. That is another reason that portable workstations are a good idea, and virtual PSAPs are possible with NG9-1-1. Demand and locations for portable workstations need to be considered. All points of failure need to be monitored, and the fact that failovers should be automatic needs to be tested. A word here about call overflow: Let's assume an area with 25 PSAPs, with one-third small (two to three) positions each, one third with medium (four to eight) positions, and the last third larger with 20–30 positions.

The failure of a large PSAP cannot be absorbed by a neighboring three-position PSAP. An overflow condition might swamp a small PSAP. A single car accident on an interstate can overwhelm a small PSAP or a large one. BCF settings can help. However, the plan for alternate routing needs to be taken seriously, with plan for failovers. The caller could get a fast busy or regular busy or announcement, but that is not what anyone wants for 9-1-1. The receiving PSAP must be able to transfer and get help from the correct first responders rapidly. NG9-1-1 was designed to enable distribution of overflow to multiple PSAPs, rather than just one. When the new NG9-1-1 network is engineered, consider many more needs than simply the core network replacement. Regulatory commissions have rules about calls in queue, overflow, and backup PSAPs. They need to be reviewed and amended or adhered to when doing design and engineering for NG9-1-1.

Inside each PSAP the equipment and computers serving each position need to be reviewed as does the cable connecting from the telco closet or IT room to the workstations. Some workstations and monitors may need to be upgraded if the same ones are planned for NG9-1-1. Operating system licenses expire, and everything needs to be compatible. That includes considering having all the PSAPs on a single IP PBX for better communications.

3.9.4 IP Address Assignment

A group of public safety agencies may share an IP engineer, or each PSAP or agency may have its own. The 9-1-1 SSP may be asked to provide IP engineers to design access from each PSAP to the ESInet and or anywhere else they need to connect. Inside each network, the IP address scheme needs to work top-down. It is a private network that needs to be integrated with monitoring to be able to identify when the location and product are not performing optimally. In the authors' experience, the 9-1-1 SSP has one dedicated IP design engineer to make all assignments and correlate them in a database to the various vendors that are part of the network. This is an especially sensitive item concerning cybersecurity, and, as a result, it needs adequate protection from intruders. The Figures 3.3 and 3.4 show IP address needs. Sometimes the interfaces to legacy equipment use SNMP protocols for monitoring. Design engineers from legacy components need to be involved to make this work smoothly and correctly

3.9.5 PC Assignment

Since SS7 is still the dominant protocol that OSPs use for accessing the 9-1-1 networks, PCs need to be assigned to the NGCSs accepting the SS7 A-links. PCs are in the form of PC XXX-XXX-XXX. Also, common language location identification (CLLI) needs to be assigned to all data centers and POI locations to be used for connectivity from the OSPs. If a 9-1-1 SSP assigns or changes SS7 providers, NGCS, database locations, or POIs, the CLLI changes and so do the SS7 PCs. iconectiv handles the assignment of CLLI and PCs. Refer to Chapter 5. The LERG is another tool that any 9-1-1 SSP must have to get their assigned 9-1-1 NGCS and other equipment locations with CLLI entered so that all connecting OSPs can prepare orders for connection of trunk facilities and links. A traditional telco design engineer can be of assistance to the 9-1-1 SSP team to get the requirements in place; assignments have upfront and, in some cases, recurring costs.

3.9.6 Domain Name System[19] and Server[20] (DNS)

A DNS is a globally distributed database for the resolution of host names to numeric IP addresses. A DNS enables queries for the names and location of different types of servers, including mail servers, name servers, and SIP proxies, The DNS allows people to use easy-to-remember text-based addresses, and the DNS translates those names into routable IP addresses, per NENA

3.10 MIS and Operations

NENA standards reference the term MIS,[21] which is an umbrella name for all monitoring and management systems including those that are part of the NGCSs such as the loggers and recorders. It must expand to include the systems that the 9-1-1 SSP must have in place to monitor the end-to-end network. Chapter 9 references monitoring and operations oversight. Chapter 11 discusses key legal and regulatory obligations for the 9-1-1 SSP.

3.11 Lessons Learned

3.11.1 OSP Carriers Providing Service

Each OSP or carrier in the footprint should be participating in meetings as needed for connection decisions, data exchange, and signaling. OSP/carrier-specific project overviews and plans should be provided to all OSPs, underlying VPC and MPC carriers, aggregators, ESInet providers, SS7, SIP, and text control center (TCC) providers.

3.11.2 Public Information

Most agencies have public-awareness programs promoting NG9-1-1. As additional functionality such as text to 9-1-1 is added, it is necessary provide public awareness and information by means of several local media outlets, public presentations and school programs, and signage.

19. DNS: A globally distributed database for the resolution of host names to numeric IP addresses per NENA. A DNS also enables queries for the names and locations of different types of servers, including mail servers, name servers, and SIP proxies. A DNS is essential to NG9-1-1 deployment.
20. DNS: Used on the Internet today to resolve domain names. The input to a DNS is a domain name (e.g., telcordia.com); the response is the IP address of the domain. The DNS allows people to use easy-to-remember text-based addresses, and the DNS translates those names into routable IP addresses per NENA.
21. MIS: A program that collects, stores, and collates data into reports enabling interpretation and evaluation of factors such as performance, trends, and traffic capacities.

3.11.3 Call Transfer Documentation

Interfaces for return of traffic for split exchanges can be voluminous and complex. Careful documentation is essential.

Selected Bibliography

NENA Detailed Functional and Interface Standards for the NENA i3 Solution.

NENA-STA-010.2-2016 (originally 08-003) DSC Approval: 08/16/2016 PRC Approval: 08/29/2016 NENA Executive Board Approval: 09/10/2016 Next Scheduled Review Date: Continuous Review; planned development as an ANSI Standard in the next version.

NENA Functional and Interface Standards for Next Generation 9-1-1 Version 1.0 (i3) NENA 08-002, Version 1.0, December 18, 2007.

NENA Emergency Services IP Network Design for NG9-1-1.

NENA 08-506, Version 1, December 14, 2011.

Development Steering Council Approval Date, November 1, 2011.

Standards Advisory Committee Approval Date, November 22, 2011.

NENA NG9-1-1 Transition Plan Considerations Information Document.

NENA-INF-008.1 (previously NENA 77-501).

DSC Approval: 07/26/2013.

PRC Approval: 10/11/2013.

NENA Executive Board Approval: 11/20/2013.

NENA Data Management for Local Exchange Carriers, ALI Service Providers & 9-1-1 Jurisdictions: Document Type: Standards; Number: 02-001 v7.1.

NENA Security for Next-Generation 9-1-1 Standard (NG-SEC) NENA 75-001, Version 1, February 6, 2010.

CHAPTER 4

Transition to NG9-1-1

4.1 Overview

The poet John Donne famously wrote in his "Meditation 17" that "no man is an island..."[1]. These are wise words for the transition to NG9-1-1. Transitioning from E9-1-1 to NG9-1-1 requires the project team to gather a significant amount of data to interconnect and interact with neighboring public safety agencies. The design of E9-1-1 is like the overall PSTN design, built as a hub and spoke network. This makes E9-1-1 public safety agencies silos that are separated from other public safety agencies and any communications with adjacent agencies (i.e., transfers) going through the legacy SR (hub) to establish a communication path. The transition to NG9-1-1 eliminates the SPOF the SR represents; the design of NG9-1-1 creates many ingress and egress points. Gathering the data necessary for a successful transition can be a voluminous and tedious task but it cannot be overemphasized that this methodical study of the public safety agencies' surrounding environment must be done to avoid project delays, design mistakes, and potentially, disaster.

Real-world experience has shown that the increased cost of NG9-1-1 causes medium and smaller public safety agencies to form regional partnerships and/or alliances to enhance service quality and reduce costs by sharing common equipment (i.e., the NGCS). Large public safety agencies and regional projects have similar issues in dealing with neighboring jurisdictions, ingress, and egress. Refer to the FCC February 19, 2016, Task Force on Optimal PSAP Architecture (TFOPA) Report, released by Timothy May.

As the NG9-1-1 project team contemplates the transition, there emerges a partnership between the public safety agencies and the award-winning 9-1-1 SSP RFP bidder and the project team assumes implementation and maintenance responsibilities. To make the transition, the two leadership teams discussed in Chapter 2 unite to meet the expectations and requirements of the customers and the public and regulatory oversight agencies. NENA provided a transition document,

1. "No man is an island, entire of itself; every man is a piece of the continent, a part of the main. If a clod be washed away by the sea, Europe is the less, as well as if a promontory were. as well as if a manor of thy friend's or of thine own were. Any man's death diminishes me, because I am involved in mankind; and therefore never send to know for whom the bell tolls; it tolls for thee." John Donne, 1623 (1624–Meditation 17).

NENA-INF-008.2-2014 called NG9-1-1 Transition Plan Considerations. This book complements those NENA standards with experience. Every project manager knows how critical documentation is. This book describes a methodology for gathering the data to build a network design. Project team members may believe that some of this data is unnecessary and that the process is trivial. In response, project management must push back and ensure that a complete, quality data set be gathered. The data is a basic building block of the NG9-1-1 transition. The knowledge to migrate to a NG9-1-1 network and system includes documentation of the "as is" emergency services network to transition to a well-designed fully redundant and resilient NG9-1-1 system.

4.2 Design Documents

Chapter 2 describes the significant amount of data shared by the public safety agencies to allow for the NG9-1-1 RFP winner to assume responsibility to become the 9-1-1 SSP. This chapter includes sample information. This additional information bolsters the timely approval and implementation of the formal applications and approvals that are critical to service quality and the financial welfare of all parties involved.

The documents needed to design and configure NG9-1-1 include but are not limited to the following: (1) jurisdiction, (2) PSAP, (3) OSPs, (4) communities served, (5) participating agencies and (6) adjacent agencies. Some material may be needed as part of the application to modify 9-1-1 service.

4.2.1 Jurisdiction

Table 4.1 may serve as a job aid. Refer to Table 4.1, which includes information to begin the data gathering and design phases, as a sample per jurisdiction.

4.2.2 PSAP Identification

To gather the information needed to assist the 9-1-1 SSP in determining the sequence of cutovers and dependencies between PSAPs, Table 4.2 may be a guideline. State clearly if a PSAP remains an E9-1-1 legacy PSAP or becomes a NG9-1-1 i3PSAP. Show the proposed number of fixed and portable[2] positions. This helps with classifying large, medium, and small PSAPs that become part of the area. With E9-1-1, the limits for alternate routing were driven by many factors. Without going into that rationale and technological limitations, understanding the current and proposed PSAP backup arrangements allows for the optimization of alternate PSAPs after they all are transitioned to the same ESInet. Any PSAP theoretically can back up any other PSAP on the ESInet network. Considerations can be weighed knowing several factors. In a perfect world a small PSAP may be alternately routed to another small PSAP, or a medium or large PSAP if capacity exists. If a large PSAP

2. Portable workstations: Heavy-duty laptops in a hard case that can be transported anywhere very easily. Since the system is IP-based, these fully functional portable workstations can connect to the ESInet and receive calls.

Table 4.1 Jurisdiction.

Jurisdiction NG9-1-1	
1. Jurisdiction: City, county, state, region, country	Enter data for each item on the left; add rows as needed.
2. Population (census)	
3. Land mass (squares mile or kilometers)	
4. 9-1-1 manager name, postal address, telephone numbers, and email address	
5. Proposed cutover date for conversion to NG9-1-1	
6. Is the agency associated formally with others in the NG9-1-1 ESInet planning? If so, name the entity, and share the name, address, contact information.	
7. Services include: voice, data, text, video, telematics, default PSAP, language specialty, social media. List all that apply.	
8. Radio frequencies and systems	
9. IGAs with adjacent agencies Y or N . Be prepared to share. (See Chapter 11.)	
10. 9-1-1 SSP name, address, telephone numbers, email	

Table 4.2 PSAP

PSAP Name and ID Number (NENA National Registry)	Number of Positions i3 or E9-1-1 PSAP	Alternate PSAP(s)	Number of Positions i3 or E9-1-1 PSAP	Add Alternate PSAP Routing Plan and Final Call Handling Treatment
PSAP 1 Name: NENA ID, full address including county, region, state, country	Fixed: Portable: i3 or E9-1-1 PSAP	Alt PSAP1 Name: NENA ID, full address including county, region, state, country	Fixed: Portable: i3 or E9-1-1 PSAP	Change or same. Plan for PSAPs then, route to least idle, or name more PSAPs. Add name of default PSAP (the PSAP if location not provided) Choose: Announcement or fast busy.

Note: Add lines or notes as needed to clarify. Add any pertinent data to assist with engineering and failover decisions (i.e., hours of operation, busy day busy hours per PSAP). If a text-only or language-only PSAP, make added notes. Any PSAP must notify the FCC at least six months ahead of turning up Text to 9-1-1. Refer to FCC.gov guidelines.

were to alternate to a small PSAP, the reverse may not work. A small PSAP, let's say of three positions, could be swamped if a PSAP with 20 positions were to send them all of its calls. NG9-1-1 routing allows for a "longest idle" PSAP designation to be used, or many other options such as those available in commercial call distributors. Some small PSAPs only operate certain hours of the day and transfer calls to a larger PSAP staffed 24 × 7 × 365. Any and all of these examples could affect bandwidth and staffing needs and may require formal changes to legal IGAs. On the ESInet, all PSAPs have access to the maps and ALI databases for the whole group of PSAPs on the network. The NG9-1-1 routing capabilities allow for routing changes that were hard to achieve in a pre-NG9-1-1 environment. Besides the sizes of PSAPs, the busy days and busy hours of each PSAP need to be taken into consideration. Then finally the 9-1-1 SSP and agencies must understand the ability for neighboring jurisdictions to dispatch for that PSAP. Radio frequencies with neighboring jurisdictions that are on the same radio systems make a difference in the ability to dispatch for the neighboring jurisdiction. One-button speed-dialing implementation of 7- or 10-digit numbers (United States) may be used to reach the neighboring jurisdiction's first responders that are not on the ESInet. Data gathering is critical to design and to build the PSAPs to handle whatever comes their way. In addition, decisions about default[3] PSAPs (primary and secondary defaults are suggested) must be assigned in case there is no ALI, and/or a call comes in without a dialable telephone number or communication device address.

Some jurisdictions allow for announcements (i.e., Intercept[4]), when all trunks to PSAPs are full or impaired while others do not. Some jurisdictions require PSAP consolidations before NG9-1-1 conversions. Some states or regions have one PSAP designated for language assistance; others have text-default PSAPs. By definition, the default PSAP accepts all calls where the caller's location identification is unknown. The default PSAP needs a secondary default PSAP. The default PSAPs require trained personnel and added resources to forward the call to the dispatch center appropriate for the person needing assistance. Common radio may be needed for default PSAPs. This is a decision for all the agencies joining together if the state or jurisdiction does not have regulatory rules.

4.2.3 OSPs and Large Enterprises

Each type of OSP and large enterprise has a separate set of issues. By gathering each OSP and enterprise by name and contact information, meetings can be held, and the necessary work can begin to assure that no 9-1-1 caller is left behind. While Table 4.3 is not comprehensive, it paves the way for planning the capacity and connections to the ESInet.

4.2.3.1 Call Location Interfaces

Transitioning to NG9-1-1 requires changes in obtaining the callers' location. Most 9-1-1 calls originate from wireless devices. Text and VoIP traffic is increasing as wireline declines. VoIP and text have similar methods of locating the caller using third parties.

In normal operation, calls that come from the wireless OSPs use MPCs to translate pANI (i.e., the area code and 555-XXXX numbers) to the handset numbers people are calling from and provide the geographic location data. Wireless OSPs associate with one or both companies currently providing MPC service. When the

---

3. A default PSAP accepts default-routed calls; Default routing is the capability to route a 9-1-1 call to a designated (default) PSAP when the incoming 9-1-1 call cannot be selectively routed due to an ANI failure or other cause. Any delays in database entry and validation may cause calls to arrive and no location to arrive with them, known as "record not found." In that instance, there are PSAPs designated as default PSAPs that the 9-1-1 SSP that receives the calls directs them to for focused assistance.
4. Announcements refer to final call-handling indicators or call intercept that advise the caller what to do next. They may be in the form of tones representing busy status or all paths busy, commonly referred to as 120 IPM. Refer to the Glossary of Terms and Acronyms.

Table 4.3 OSP and Enterprise per Jurisdiction

Jurisdiction Name:

Carriers (OSPs)		*Street Address, City, Zip Code*	*Telephone Number*
ILEC/RLEC			
ILEC	Name 1	Contact name and reach information; dedicated transport-and signaling	
RLEC	Name 2	Contact name and reach information; dedicated transport and signaling	
CLEC			
CLEC bandwidth.com VPC	Name 1 using aggregator	Contact name and reach information; add signaling	
CLEC VoIP and wireless transport	Level 3 communications aggregator	Contact name and reach information; common transport provider; add signaling	
CLEC West VPC for database	Name 2; dedicated facilities	Contact name and reach information; dedicated transport; add signaling	
VoIP			
VoIP West and Comtech VPC	Name 1 using aggregator	Contact name and reach information; common transport provider; add signaling	
VoIP West VPC	Name 2 using aggregator	Contact name and reach information; common transport provider; add signaling	
VoIP Bandwidth VPC	Name 2 using aggregator	Contact name and reach information; common transport provider; add signaling	
Wireless			
Wireless West MPC	Name Mobility w Text	Contact name and reach information; add signaling	
Wireless West MPC	Name PCS w Text (SIP using Level 3 SIP transport liaison says Level 3 selects signaling)	Contact name and reach information; add signaling	
Wireless West MPC	Name cellular w text	Contact name and reach information; add signaling	
Wireless West MPC	Name w Text (using Name 3rd party Transport)	Contact name and reach information; common transport provider; add signaling	
MPC/VPC E2 Data Links into each Data Center			
VPC	Bandwidth	Contact name and reach information	
MPC/VPC	Comtech TCS	Contact name and reach information	
MPC/VPC	West	Contact name and reach information	
Reseller			
Reseller	Name via which Major OSP	Contact name and reach information; common transport provider; add signaling; when you test their carrier only, they can set up access through testing	
Text Control Center			
TCC	Comtech TCS or others offering service	Contact name and reach information; common transport provider; add signaling	
Carriers not on diagrams because: Out of Business, no Customers, not on active state commission list			
Uncertain category	On list in LERG or diagram from E9-1-1; not able to identify active status	Contact name and reach information Some are difficult to locate; working with the agencies, regulatory bodies, and the carrier list on NENA, plus asking the prior 9-1-1 SSP are options	
Enterprise Major Customer (from Legacy SR to connect via ESInet to NGCS to PSAP)			
Name	Contracted with Company Today	Identify contacts, and signaling today. Separate contracts are required for each private entity. The legacy SRs cannot be disconnected until this step ends. Each state or jurisdiction has rules about enterprise connectivity. Often enterprises are early SIP signaling adopters with IP PBXs.	

geographic location data arrives, the NG9-1-1 system knows which PSAP to route calls to. If critical location and related ANI does not arrive, the calls default-route to a designated primary or secondary PSAP where the call is answered, and the call takers try to determine the right PSAP and transfer the call. In a default-route occurrence, a trouble ticket should be created and the technical team is notified to fix the problem. It could be a technical problem like garbled call setup or that the two diverse E2 datalinks[5] are both down or that the two diverse databases are down. It could be a failure of the database to return an accurate caller number or location.

VPCs are similar to MPCs. VoIP OSPs use one of the currently three companies to provide similar functions as the MPCs. On VoIP calls the pANI is recorded as NPA 222 XXXX. Fixed-VoIP services' locations are typically accurate by the very nature of the service being at a fixed location. The issue with VoIP comes with mobile (nonfixed) VoIP service where the user is required to provide the address location that is provided to the 9-1-1 system for routing. Failing to update the location causes the call to be misrouted. There are stories of overseas 9-1-1 calls being received at a U.S. PSAP. This is an inherent flaw in the service and depends on the general public to grasp the basic concepts of 9-1-1. "No one ever went broke underestimating the intelligence of the American public."[6] The agency director and 9-1-1 SSP need to be aware of the VoIP location issue and include in the planning and testing process.

As text to 9-1-1 becomes more prevalent, TCCs have formed and function similarly to MPCs and VPCs, but there are options. Each 9-1-1 SSP must choose one of the three NENA options for connectivity. Currently the three options NENA provides (reviewed in Chapter 10) are all interim solutions. Each OSP with text-to-9-1-1 capabilities interacts with the TCC and the 9-1-1 SSP to make the actual texts come into the network and process correctly.

In the transition, PSAP-connected e2 datalinks to the MPC, VPC, and TCC move to each NGCS at the diverse data centers to create redundancy and resiliency.

4.2.3.2 ALI MSAG Transition

In the transition to NG9-1-1, location information formerly in the MSAG—now the LVF—may require additional data fields to allow the ECRF to make more precise routing decisions by adding more specific data to each record such as the community and state field. Some public safety agencies may push back on having to update the state field, despite the fact that it is useful today in making the records more unique and that it will only become more useful in the future as ESInets are joined.

Along with using the additional fields in the NENA standard ALI record that the MSAG uses to validate the ALI record, the ALI record transition involves geocoding each ALI record address with X/Y coordinates. The 911 SSP must coordinate with the GIS, map, ALI/LIS database, and ECRF vendor(s) as to the specific fields that are included. This information must be communicated to the OSPs providing ALI/LIS records.

---

5. NENA-05-001, December 2003, NENA Standard for the Implementation of the Wireless Emergency Service Protocol E2 Interface.
6. Henry Louis Mencken.

The GIS data preparation can be the longest and most time-consuming task in the project. If the public safety agency has complete, updated, and edited GIS records, it is ahead. One of the main advantages to NG9-1-1 is the location-based routing, and GIS makes that magic happen.

During transition, weeks before the scheduled cutover date, duplicate updates are made to the respective ALI and LIS databases. The new NG9-1-1 landline database process improves reliability in routing the call. See Chapters 8 and 10.

Table 4.5 is an example of the lists to be gathered for public safety agencies that directly support a jurisdiction preparing for NG9-1-1. This information is essential to building the NG9-1-1 databases for routing and contacts essential for each PSAP to be able to reach the proper first responders and dispatchers depending on the caller's requirements. New databases and maps must work synchronously on the same ESInet and between ESInets. Note that this book does not go into detail about the lengthy process of building the GIS and NG9-1-1 databases. NENA has many standards documents regarding database management. Refer to NENA STA-015.10-2018; NENA Standard Data Formats for E9 1 1 Data Exchange & GIS Mapping; NENA-STA-006.1.1-2020; NENA Standard for NG9-1-1 GIS Data Model (also includes NENA-REF-006.1-2020 and NG9-1-1 GIS Template Files); and NENA-STA-012.2-2017, NG9-1-1 Additional Data Standard. There are more database-related documents. Refer to NENA.org. This book assumes that the NENA database standards are understood for all types of services including wireline, wireless, and VoIP services and telematics. The subject of database management and GIS could be a technical book of its own.

4.2.3.3 Direct Connection

NG9-1-1 gains reliability and diversity by direct connecting with the OSPs. This transitions the OSPs from terminating their 9-1-1 traffic into an aging SR. While an OSP may provide service into the geographic area served by one Public Safety agency for call completion in a county, they may not serve all counties with PSAPs on the ESInet. PSAP at a time research is required. If an ESInet serves fifteen (15) counties and thirty-five (35) PSAPs as many as ~hundred or more OSPs may connect to anywhere from those one (1) to fifteen (15) counties.

4.2.3.4 Selective Router and Split Exchanges

E9-1-1 SSPs

Each E9-1-1 SR in the footprint must be identified and associated with the OSPs and PSAPs. In phased transitions, the legacy SR remains connected and/or needs to be used for an interim period to return calls for split exchanges. Chapter 5 explains split exchanges and contains OSP information and Figures 5.6–5.8. Enterprises (detailed in Chapter 6) require similar if not more complex solutions to migrate to NG9-1-1 in comparison with the average OSP. Identifying connected enterprises has proven to be most challenging in real-world conversions; see Table 4.3, which can be expanded as needed.

4.2.4 Communities Served

Part of the data gathering process for areas choosing to convert to NG9-1-1 incudes identifying all the local communities served. This is important, because it is necessary to identify boundaries for routing callers and to identify all the first responder types who serve an area. Some communities have many agencies providing fire and ambulance protection. In our experience, we found one town with few residents over 30-plus square miles being served by five different fire protection districts, with public and private ambulance services, and one police department. Boundaries differ significantly around the world. This may seem a simple step, but it is an essential one leading to the next steps of identifying the participating first responder agencies (Section 4.2.5) and adjacent first responder agencies (Section 4.2.6). See Table 4.4.

4.2.5 Participating Agencies

Table 4.5 is an example of the lists to be gathered for public safety agencies that directly support a jurisdiction preparing for NG9-1-1. This information is essential to building the NG9-1-1 databases for routing. Table 4.5 also provides contacts essential for each PSAP to be able to reach the proper dispatchers and first responders. In all cases, new databases and maps must work synchronously on the same ESInet.

To use Table 4.5, provide a list of public safety agencies (e.g., police, fire, EMS, and ambulance.) that are to be dispatched by the NG9-1-1 system, including each agency's land area in square miles and estimated population that has access to the proposed NG9-1-1 system. Do not forget to include sheriff and/or state police jurisdiction and state/province/parish police districts. Each agency that appears on this list should also have signed a call-handling agreement (i.e., IGA); refer to Chapter 11. Next, extract a summary of all listed communities and compare to communities served. Recognize that there may be agencies serving next-door PSAPs in a different county or under differing jurisdictions.

4.2.6 Adjacent Agencies

Adjacent agencies may already become part of an existing ESInet the day this agency applies or in the future when the remaining agencies come on board. Sequence of actual cutover determines the status. Where possible, be as specific as possible if adjacent counties or agencies are either part of an ESInet already or not, the same ESInet, or another ESInet. The adjacent agency could be on an ESInet belonging to another 9-1-1 SSP in same state or even an adjacent state or country. This is going

Table 4.4 Communities Served

Communities Served	
Public Safety Jurisdiction Name:	
Provide a list of all communities to be served by the proposed NG9-1-1 system. Include the name of the community and the official mailing address including street address, city, and zip code, and the source such as individual community websites	
City, Town, or Village	*Street Address, City, Zip Code*
Name	Address

Table 4.5 Participating Agencies

Participating Agencies							
Public Safety Jurisdiction Name:							
PSAP Name(s)							
Instructions							
NG9-1-1 Participant Agencies	*Street Address, City, Zip Code*	*Administrative* Telephone No./ Contact*	*Direct Dispatch*	*Transfer*	*Call Relay*	*Agency Land area(s) in square miles*	*Estimated population*
Name County Sheriff's Dept. (Name County)	Address usable by GIS Database	Number to be entered into the NG9-1-1 system for one-button transfer with in-charge person and telephone, and other reach	Radio frequency	Enter method or X if No option	Enter method or X if no option	Obtain and enter per GIS resource	Typically census; population of area varies per agency
Community 1 Name 1 Fire Department Name							
Community 1 Police Dept. (Name Department)							
Community 1 Ambulance Service Name							
Summary	#Countries	# States	#Counties	#Agencies	# State Police		

*Administrative numbers are a term-of art-used in this document. The authors are fully aware that the goal is to have all calls and transfers between PSAPs be handled by ESInet trunking. Some public safety agencies have dedicated 24×7 emergeny7/10-digit lines for transfer; because smaller PSAPs may not have dedicated lines, they use any lines available. The i3PSAPs have the capabilities built into them along with IP PBXs among agencies to transfer calls seamlessly around the ESInet and off the ESInet as needed.

to take research. States may have websites showing approved applications. State NENA groups of leaders know each other and often know if a neighbor or agency is cutover or pending a cutover. In some cases, a county or jurisdiction may have purchased and cut over their equipment to be IP-enabled (NG9-1 1 compliant) but not have changed to a NG9-1-1 9-1-1 SSP configuration. To the world they get all calls through a legacy SR, and no one else was involved. They may simply have needed an upgrade and bought IP-enabled NG9-1-1 equipment to be ready. Considerations to invite such an adjacent agency to join an ESInet may depend on financial arrangements or compatibility testing. Completing this information and ensuring that it is accurate is going to take time and almost always involves IGAs between neighbors. Upgrades involve planned testing with adjacent agencies ahead of cutover and live testing during the actual cutover to be sure that everything is operational. These records (all the forms shown in Tables 4.1–4.6) are not static. They require constant upgrades, including database upgrades to the NG9-1-1 systems and routing instructions.

Instructions to complete the form in Table 4.6 may read as follows: Provide a list of public safety agencies and existing 9-1-1 systems that are adjacent to the proposed system's boundaries. Each agency that appears on this list should also have signed a call-handling agreement and/or aid outside jurisdictional boundaries.

4.3 Transfers On-net and Off-net

There are many reasons calls must be transferred between i3PSAPs and E9-1-1 and as the PSAPs move to NG9-1-1 network architectures. The simplest transfers occur when PSAPs are on the same ESInet. It takes time to get all the PSAPs moved (and if faced with regulatory oversight, approved) to NG9-1-1 and to move one agency at a time to an ESInet, but the project must begin somewhere. The agency-by-agency

Table 4.6 Adjacent Agencies

Adjacent Agencies *Public Safety Jurisdiction Name:* *PSAP Name(s)* *Instructions* *Home State*			
1. Name sheriff's department; name adjacent county 1	Full street address, zip code	Telephone number, Sheriff name	Means of transfer of call: telephone 10D fixed transfer 10-digit telephone number; and show date when transfer is via ESInet or future ESInet to ESInet
2. Town area ambulance, adjacent county1	Full street address, zip code	Telephone number name, coordinator	Telephone 10D fixed transfer telephone; via ESInet date
3. Town fire department, adjacent county1	Full street address, zip code	Telephone number fire chief name	Telephone 10D fixed transfer telephone; via ESInet date
4. Name county sheriff's department, name adjacent county 2	Full street address, zip code	Telephone number Sheriff name	Telephone 10D fixed transfer via ESInet at cutover
5. County sheriff's department, Adjacent county 3	Full street address, zip code	Telephone number Sheriff name	Telephone 10D fixed Transfer telephone; via ESInet date
Add to list			
Adjacent State			
Name fire and rescue name county state	Full street address, zip code	Telephone name, chief	Telephone 10D fixed transfer telephone; TBD transfer interstate ESInet to ESInet
Name fire and EMS name county state	Full street address, zip code	Telephone number fire chief name	Telephone 10D fixed transfer telephone; TBD transfer interstate ESInet to ESInet
Add to list			
n Country no police agency transfers	n Home state States n Home ESInet Counties n Home ESInet agencies	n Adjacent state n Adjacent ESInet to ESInet Counties n Adjacent ESInet-to-ESInet agencies	No trunk Transfers—Future ESInet within Home ESInet Cutover; Interstate ESInet Transfers are possible as adjacent state expands the NG9-1-1 footprint.

total transition period from start to finish ideally takes no more than a year, because it is expensive to operate two networks and keep changing network interfaces, testing, and retesting. Since there is seldom a statewide "grand design," the 9-1-1 SSP and the agencies that join the new ESInet need to have a solid executable plan. By now the data has been gathered, and a sequence is documented for the project team, and shared with all of the OSPs, major enterprises customers and neighboring public safety agencies.

PSAPs connected to the ESInet become on-net PSAPs. When the first agencies are approved and cut over, all other agencies become immediately off-net, and transfers to all off-net agencies must be facilitated. This is not simply a data-gathering exercise; it becomes the basis of trunking and routing interfaces that must be built, even if temporarily, to reach agencies previously connected with legacy SR trunking. Chapter 7 addresses this new trunking and routing design. Chapter 8 covers each stage of the transition to NG9-1-1 to be sure transfers work. Finally, this data gathering helps to determine the sequence of the cutover, as agencies are migrated from E9-1-1 to NG9-1-1 ESInet networks. See Table 4.5.

PSAPs need to be moved on-net, ideally within same eight- to 24-hour period. We recommend the first agency cutover to have at least two PSAPs that can back each other up. Chapter 8 covers this topic in more detail.

4.3.1 PSAP Transfers

PSAPs have many needs for transferring calls. Often the biggest reason for transfer is that the caller is moving and crosses a PSAP boundary, as is often the case with wireless callers. Transferring might also be necessary when a call comes to a PSAP without the location information needed to find the caller and determine the nearest PSAP that should receive the calls. In that case if the NGCS is designed as suggested, the call first routes to the default PSAP, where additional information is obtained if possible, by a voice or text response request, "9-1-1 What is your Emergency," followed by "What is the Location of the Emergency." Assuming no reply, the default PSAP may have the NPA NXX of the caller if it comes from a wireline OSP, and it should have a table accessible through the PSAP workstations of the communities associated with the NPA and NXX. They determine the most likely PSAP and do a warm handoff to the agency matching that combination. It is not a guaranteed match, because of local number portability (LNP) and the propensity for CLECs to have a single NPA NXX serve a large overlay area. With wireless and VoIP OSPs at least an NPA 555 or NPA 222 and number series comes up identifying the OSP assigned to that range of pseudo ANI numbers. The call taker tries everything possible if a caller can speak but has no idea where they are located. A photo of a nearby room location or street may be requested. As mentioned, many times, 9-1-1 location services may not be optimal, but the caller may have a location app on their own phone that might be helpful through social media or other connections or apps. Another reason for transfer may include if a text-only PSAP or a language-only PSAP is predesignated to transfer those callers to a PSAP dedicated to helping callers with those special needs. The PSAP that receives a call and determines that it does not belong in its PSAP does all it can to transfer the caller to the correct agency/PSAP capable of dispatch and stay with the caller until help arrives.

Transfers can occur by PSAP to PSAP on-net via the NGCS ESRP, or PSAP to PSAP off-net via the same NGCS ESRP. The ESRP sends the calls out to trunks connected to the E9-1-1 legacy SR via the LSRG to reach the off-net PSAP, or even dial a seven- or 10-digit number for a transfer. The concept of one-button transfer means that the speed dial for the agency is built into the console for the call taker to avoid a lookup or misdial. If a PSAP establishes direct trunking to another PSAP, the ESRP to LSRG to those trunks would be utilized. This depends on the network and how things are designed in the first place. If the PSAP on the ESInet is still a legacy PSAP not converted to i3PSAP, LPG) trunks are involved. i3PSAPs can always forward all caller's information to another i3PSAP as long as SIP signaling is not interrupted whereas E9-1-1 PSAPs do not have SIP interfaces. Chapter 5 discusses call routing back through ESRPs to LSRGs through the old E9-1-1 SR for split exchanges. They may be needed for other transfer reasons. All these scenarios must be tested. Sometimes 7/10-digit transfers may be the method of last resort, the only option, until ESInets are connected to other ESInets and everyone nearby has been converted to NG9-1-1.

After the project team reviews the participating and adjacent agencies, it documents the transfer plan and makes sure it works. Note that when trunking is set up between agencies, it takes capacity. Be sure when engineering interfaces that there are sufficient trunks available for those voice and text transfers. A lot of transfers eventually become ESInet-to-ESInet calls, but not at first. We have heard of blocked calls occurring because this was not taken into consideration.

4.4 Communications Alternatives

Here we consider some simple alternatives for PSAP-to-customer and PSAP-to-PSAP communications.

4.4.1 IP Phones

As part of an RFP, the bid usually includes the addition of a new IP PBX for use among all the agencies joined together over the ESInet, including directors, managers, and PSAP users. Everyone they may need to reach expeditiously can be connected through extensions of this new IP PBX integrated with the i3PSAPs. Assignment and integration of these PBX lines must be managed and tested. Not all VoIP connections are equal in terms of quality. Get a good recommendation and a commitment for quality of service (QoS) with a maintenance agreement.

4.4.2 One Button

One-button transfers is simply a built-in speed-call arrangement for participating and adjacent agencies and the most critical numbers that must be reached if an on-net transfer is not possible or fails for any reason. This may be the first method of communication.

Transfers to other numbers are another consideration. There are many N-1-1 numbers and the Suicide Hotline with its new national number approved in the United States, 9-8-8.[7] When calls arrive at the PSAP, there are transfer and call-handling processes that should be consistent across the PSAPs sharing an ESInet. If any PSAP can receive any call, a decision must be made to reject the call (i.e., tell the caller to use the appropriate agency numbering plan, such as 3-1-1 for nonemergency numbers) or to attempt a call transfer to the proper agency such as a call for the Suicide Hotline, which has the three-digit number 9-8-8 in the United States. If a transfer of any special number is mandated or needed, the PSAP transfer solutions need to have either trunking or a means to automatically extend the call for assistance and stay involved until the caller is securely connected. Unique situations must be documented in the planning stage, with the network set up to make the plan work, and then included in testing, cutover plans, and the PSAP training.

4.4.3 Radio Transfers

The data gathering includes radio transfers for all the ESInet-participating agencies. Consider what radio transfers are possible as an alternative for adjacent agencies. It may be helpful to research emerging FirstNet capabilities to reach dispatch agencies and PSAPs via radio. At some point NG9-1-1 will interface with FirstNet. Plan ahead and stay connected to the FirstNet working teams where authorized and funded to optimize the results of that investment to save lives. From prior work, many agencies joining an ESInet already had a common radio solution, Starcom,[8] which allowed them to take calls for each other and still be able to transfer back for dispatch to agencies several counties away from them seamlessly. The NG9-1-1 network in the CSI used common radio at least five years without blocked NG9-1-1 calls.

4.4.4 Satellite Phones

There are instances where a PSAP agency uses satellite phone backup capabilities including satellite trucks as backup for major events or disaster recovery scenarios when all other communications fail. This is not connected to NG9-1-1 or the ESInet and is mentioned here for consideration when "all else fails." Prudent emergency services managers plan ahead and use all means of communications to

7. https://docs.fcc.gov/public/attachments/FCC-20-100A1.Exerpt as follows: "FCC Staff Report recommended that the Commission initiate a rulemaking to designate a 3-digit dialing code for a national suicide prevention and mental health crisis hotline system, and that the Commission consider designating 988 as the dialing code for this important purpose in this Report and Order, we designate 988 as the 3-digit number for the Lifeline. We also address implementation of 988 in detail. In particular, based on the record, we require all covered providers to fully implement 988 in their networks by July 16, 2022."The FCC Approved their Report and Order in July 2020. Implications for those calling 9-1-1 who need immediate assistance are not specified." Refer to Chapters 4, 7 and 8 for transfers or dispatch. Calls are forwarded by each OSP and Large Enterprise to (800) 273-8255 USA.
8. "STARCOM21 is a public/private partnership with Motorola Solutions envisioned and commissioned by the state to enable seamless, interoperable communications among state, local and federal government users. The network primarily focuses on the daily, two-way radio communication needs of its subscribers, enables enhanced response to disaster and emergency situations, and allows the state to effectively address homeland security concerns." This is used as an example not an endorsement.

supplement their strategies and reach out to state and federal law enforcement in times of localized and national challenges.

4.5 ESInet On-net to Off-net PSAPs

Figure 4.1 shows an example of a new ESInet with a single agency and multiple PSAPs residing on the ESInet while the remaining agencies are awaiting testing and conversion to NG9-1-1. In Figure 4.1, a wireless caller enters 9-1-1 into their smart phone reaching their designated PSAP, which has been converted to NG9-1-1; by the time the call is received, the NG9-1-1 PSAP realizes that the neighboring PSAP should have the call because the caller has moved to their jurisdiction, and the PSAP that has the optimal dispatch capabilities is in a PSAP that has been converted to an i3PSAP but remains behind a legacy SR. The first PSAP that receives the call evaluates the real-time information and initiates a call transfer. The call is returned to the NGCS in the data center where routing directs the call via trunking interfaces to the legacy SR that is connected to the neighboring PSAP. The live handoff is made, and the appropriate first responders are dispatched to care for the caller. The

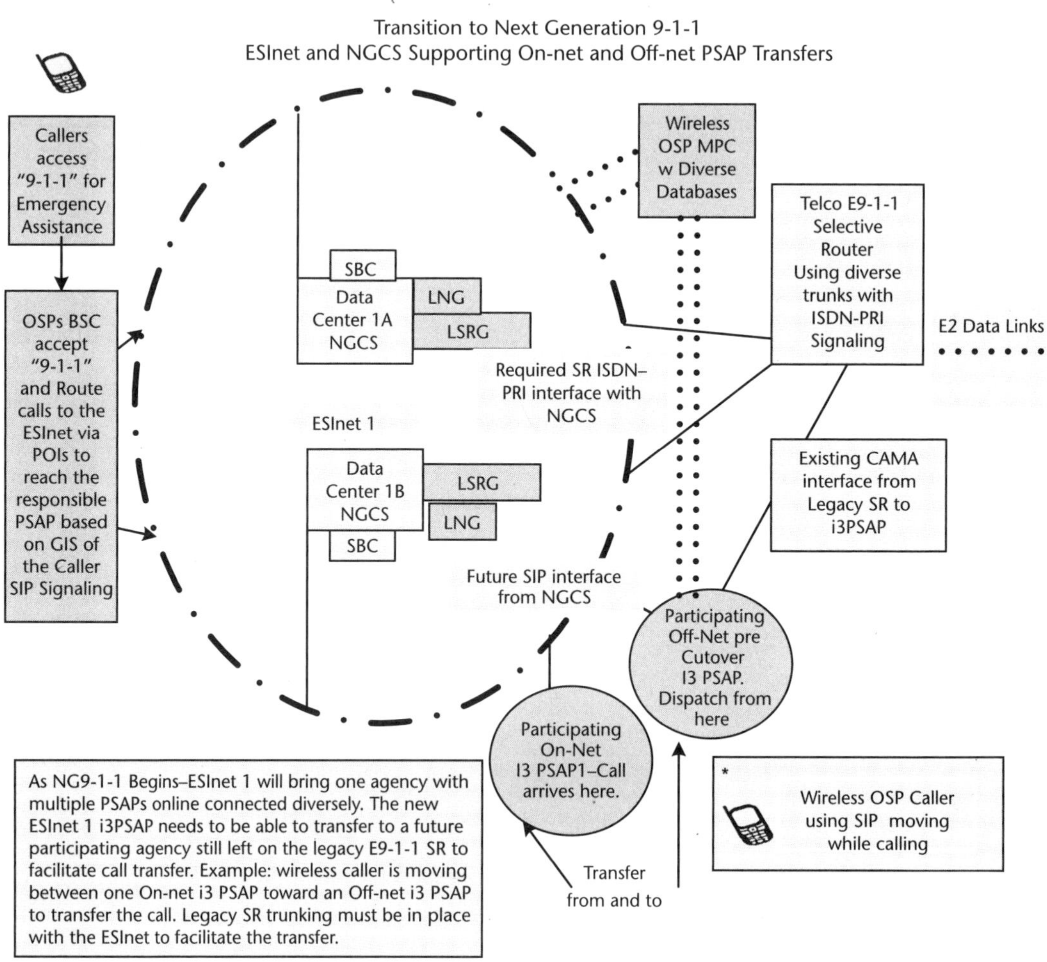

Figure 4.1 ESInet On-net and Off-net.

initial PSAP is an on-net PSAP (i.e., on the ESInet) as a participating agency. The destination PSAP become a participating on-net PSAP once it is cut over. For now, it is a participating off-net PSAP. There was an option for (1) trunk transfer, (2) a 10-digit number transfer, or (3) a radio transfer if the PSAPs use compatible radio systems. If the legacy SR had not been connected to the ESInet, the trunk transfer option would not have been viable. Again, the data gathering allowed scenarios to be constructed, and the NG9-1-1 systems to be built, tested, cut over, and managed through each succeeding cutover.

4.6 ESInet to ESInets

Figure 4.2 demonstrates an ESInet-to-ESInet scenario where another wireless caller requires 9-1-1 assistance. The caller again is routed to ESInet 1, and the call is received at another participating on-net i3 PSAP. In this scenario, the adjacent agency is not converted to NG9-1-1, but it resides on an ESInet. The LPG handles the conversion to the E9-1-1 PSAP. The call is routed between the two ESInets and to the E9-1-1 PSAP. Because the adjacent PSAP is not i3-compatible, the first PSAP call taker must do a warm handoff and share all the caller information. The adjacent off-net PSAP dispatches to the appropriate first responder. Eventually the E9-1-1

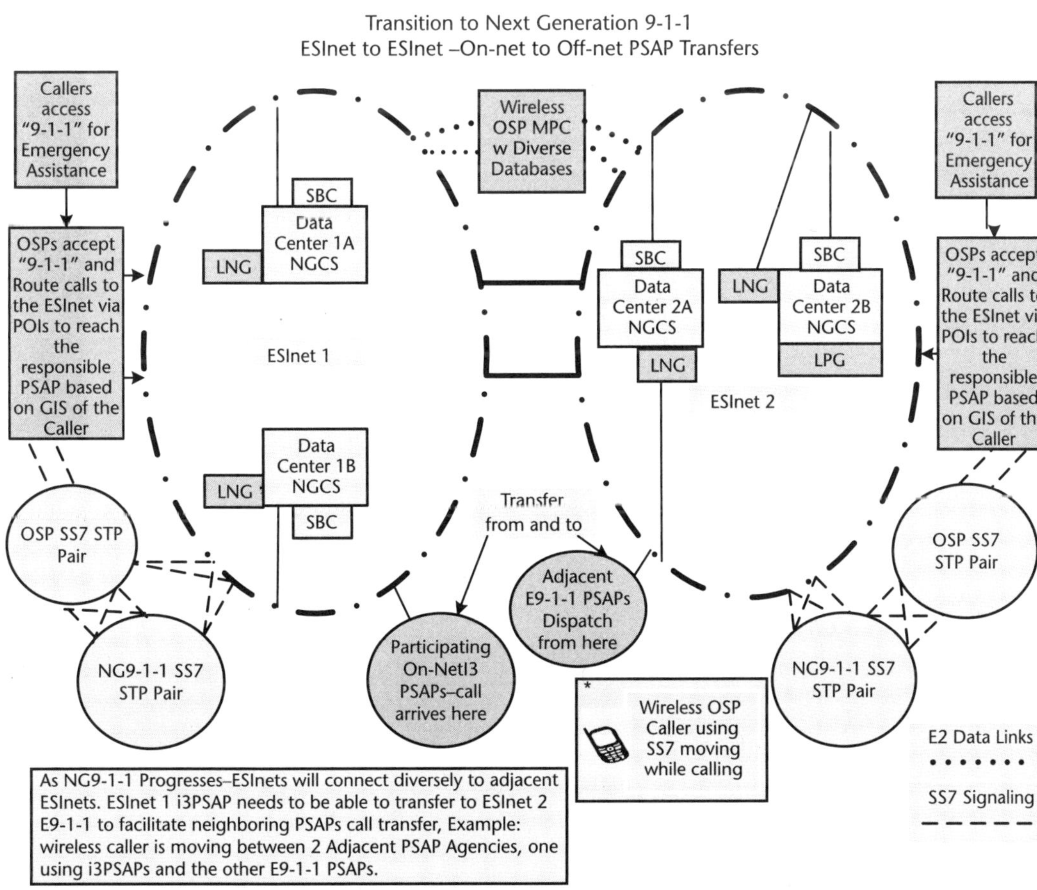

Figure 4.2 ESInet to ESInet–PSAP Transfers.

PSAP gets converted, and all information flows to the second PSAP even though they are possibly in another county, state, or theoretically, country. These examples are certainly not comprehensive. They set the stage for variables that must be used in testing. In this scenario the options were the same as before: (1) trunk transfer, (2) seven–10-digit transfer, or radio transfer if compatible. Signaling variables are used randomly in each example. The PSAPs do not need to be aware how the call is progressing, or how and where they get ALI and ANI. They simply need to receive the call and the ANI and ALI information to do their job. The 9-1-1 SSP and agency managers need to ensure that all possible diversity and options are built correctly and work each time. This book's extensive figures aim to allow readers to view technical options. Chapter 7 lays out a rigorous process for testing the new NG9-1-1 systems and network; the new PSAPs; the access OSPs; and enterprises, signaling, and interfaces to legacy SRs and SS7 network components, databases, and security.

NENA documents provide options in case an ESInet fiber ring or broadband connectivity is not available to an area. NENA standards allow for many configurations, and many types of ESInets. There are options for statewide ESInets that have regional ESInets connected to them. NENA deals extensively with the reliability and diversity requirements to reach optimal 99.999% reliability for ESInets and 99.995% reliability in the PSAPs. Our experience has shown that there are cost benefit tradeoffs and that a highly reliable network is achievable if the design is well considered end-to-end.

Selected Bibliography

NENA-INF-008.2-2014 (originally 77-501), NG9-1-1 Transition Plan Considerations.

NENA-05-001 December 2003 NENA Standard for the Implementation of the Wireless Emergency Service Protocol E2 Interface.

CHAPTER 5

Originating Service Providers

This chapter addresses OSP roles and responsibilities implementing NG9-1-1. The terms carrier, access carrier, and OSP are often used interchangeably within standards, industry documents, and legal and regulatory filings. OSP is the "term of art" used in this book. This chapter covers OSP requirements and options, as they plan to connect their networks to NG9-1-1. OSPs include: wireline, wireless, and VoIP service providers. Guidelines and standards govern how each type of OSP may interconnect for NG9-1-1. The standards allow for each OSP to choose its trunking, signaling methodology, underlying transport connectivity, and diversity.

5.1 OSP Overview

5.1.1 Wireline OSPs

In the United States, there are distinct types of wireline OSPs: ILECs, regional local exchange carriers (RLECs), and CLECs. Table 5.1 summarizes the differences among these OSPs. ILECs and RLECs offer voice, TTY for persons with disabilities, and multiple customer classes of service from residential to business. The difference between ILECs and RLECs derives from their historical role as the monopoly providers of switched telephone service. ILECs and RLECs are regulated by the FCC and state regulatory agencies. An ILEC cannot route customer calls directly across a LATA boundary for completion. ILECs must hand off interLATA calls to an interexchange carrier (long-distance carrier). Over time there have been consolidations, mergers, and acquisitions. Regulations have been modified from the time of the Modification of Final Judgment (MFJ)[1] but the 9-1-1 mandates remain.

1. In U. S. telecommunication law, the MFJ is the August 1982 agreement approved by the court (consent decree) settling *United States v. AT&T*, a landmark antitrust suit, originally filed on January 14, 1949, that modified the previous MFJ of January 24, 1956. Entered into between the U.S. Department of Justice and the American Telephone & Telegraph Company (AT&T), the MFJ, after modification and upon approval of the U.S. District Court for the District of Columbia, required the Bell System divestiture of the Bell Operating Companies from AT&T. Judge Greene was the presiding judge for the case. The exact term for the MFJ is given under the general definition of LATA, a term used in U.S. telecommunications regulation for geographic areas in which local telephone companies are allowed to provide service; the term being used is "the Modification of Final Judgment (MFJ) entered by the United States District Court for the District of Columbia in Civil Action number 82-0192 or any other geographic area designated as a Local access and transport area (LATA) in the National Exchange Carrier Association, Inc. Tariff FCC No. 4." The Bell System was the system of companies, led by the Bell Telephone Company and later by AT&T, that

Table 5.1 Wireline OSPs

Type of OSP: Wireline			
Category	*Examples*	*Service*	*Attributes*
ILEC	AT&T, CenturyLink, Verizon, Frontier	Voice, TTY for persons with disabilities; LNP; multiple classes of service	split-exchange routing by the SR function; fixed location address; direct connection to NGCS.
RLEC	Big River, Egyptian Telephone, Shawnee Telephone, TDS Telecom (RLECs are numerous.)	Voice, TTY for persons with disabilities; LNP, multiple classes of service	Split-exchange routing by the SR function; fixed location address; direct connection to NGCS.
CLEC	Clearwave, TDS Metrocom, CenturyLink, Comcast; dozens of OSPs including ILECs and RLECs owning CLECs outside their original footprint	Voice, TTY for persons with disabilities, LNP, multiple classes of service	7–10-digit routing for exchanges to reach the proper PSAP from originating switch; fixed location address in MSAG—preferred or via third-party aggregator
Note: Enterprise—with direct access by passing the local OSP for 9-1-1 service	Large businesses, campus environment, manufacturing facilities, government entities, hospitals, and universities	Enterprise voice, TTY for persons with disabilities, LNP and multiple lines of service	Refer to Chapter 6

RLECs (sometimes known as independents or "mom-and-pops") are those companies whose businesses operate in a fixed geographic area contiguous with the former Bell system companies and that perform similar functions for their customers. At one time there were approximately 1,400 independent companies. Over time, there have been acquisitions and consolidations within this sector; however, the RLEC designation continues to apply. RLECs are not subject to the interLATA restrictions of the MFJ.

CLECs[2] are the third wireline OSP category. There are two broad categories of CLECs, facility-based and resellers. Since resellers do not own any part of the network and rely on the underlying OSP for all signaling, transport, and routing, this book discusses only facility-based CLECs. Often the CLEC owns a switch, which could be either TDM or IP, and purchases the transport and last mile from other OSPs. Aside from this, the services CLECs provide are similar to ILECs and RLECs. Their method for routing emergency service calls is different, however, and it is important to understand those differences. CLECs often use LNP.[3] This ability of customers to move telephone numbers between OSPs means that the CLEC may

provided telephone services to much of the United States and Canada from 1877 to 1984, at various times as a monopoly. On December 31, 1983, the system was divided into independent companies by a U.S. Justice Department mandate.

2. A CLEC, in the United States and Canada, is a telecommunications provider company (sometimes called a "carrier") competing with other, already established carriers, generally the ILEC.
3. The Telecommunications Act of 1996 [47 U.S.C. § 251(b)(2)] established that all local exchange carriers must implement LNP in accordance with the regulations of the FCC. March 26, 2015, the FCC approved the recommendation of the North American Numbering Council (NANC) to award the contract to Telcordia Technologies, doing business as iconectiv, as the next LNP administrator (LNPA). The iconectiv contract was finalized in August 2016. iconectiv officially became the administrator of the NPAC on May 25, 2018.

have lines that appear to belong to an ILEC or RLEC. CLECs can apply for NXX[4] codes, or exchange numbers unique to their COs within their boundaries. Split exchanges, detailed in Section 5.11, are not allowed for most CLECs. CLECs use the full seven- or 10-digit number analysis at their own switching locations before routing calls to the 9-1-1 SSPs to any number of PSAPs to resolve their split exchange–like conditions requiring many 9-1-1 dedicated trunk groups from a single CLEC CO. Often a single originating switch for a CLEC covers a whole state. CLECs may use third-party aggregators for location services even though their customers have fixed address locations. There are hundreds of CLECs in business across the United States. Third-party providers may aggregate many OSPs' calls to deliver them to the designated 9-1-1 SSPs. Locating all the CLECs in the 9-1-1 SSP area can be a challenge since their COs may be outside state or local boundaries. Many of the CLECs have soft switch technology and can send their 9-1-1 calls using SIP signaling, which is ideal for NG9-1-1.

A few additional observations and distinctions between the three types of wireline OSPs are summarized as follows:

1. An ILEC or RLEC can own a CLEC business outside of its own geographic exchange boundaries, or a CLEC can be a totally new local access competitive carrier with no ILEC or RLEC interests.
2. ILECs and RLECs are mandated to offer LNP to customers who choose to keep their telephone numbers when they move their service to wireless OSPs (cellular carriers), CLEC OSPs, VoIP service providers, and resellers.
3. ILECs and RLECs may have split-exchange routing for emergency services performed at the legacy SR tandem switch function rather than doing further translations in each end office for a split exchange.

5.1.2 Wireless OSPs

Wireless OSPs provide voice, data, and text service for their customers and provide the gateway for wireless services (i.e., telematics,[5] emergency services calls generated automatically by vehicles, and the IoT), calls generated by sensor networks or arrays. Wireless OSPs, including those that provide personal communications services (PCS), account for approximately 80% of calls to 9-1-1 in the United States. Text messages originating from wireless OSP networks can be delivered to the 9-1-1 SSP. Chapters 1 and 10 address text to 9-1-1.

There are two types of wireless OSPs, distinguished by the methods they use to access NG9-1-1 networks. One type of wireless OSP uses direct access to connect from their MSCs to NG9-1-1 networks. The other type delegates the access methodology to third-party MPCs or transport providers. Table 5.2 summarizes the differences between these two types of wireless OSPs.

4. NXX refers to the three-digit telephone exchange code assigned by NANC where N is any number two–nine and X is any number zero (zero–nine.
5. At its simplest, telematics is a way to collect and use vehicle data to achieve any number of tasks, such as locating a vehicle and other monitoring data using signals from GPS telematics. Today, companies can collect a wide variety of telematics data. Emergency services requests are relayed through the wireless OSP networks to their proper designation.

Table 5.2 Wireless OSPs

Type of OSP: Wireless			
Category	*Examples*	*Service*	*Attributes*
Wireless—direct-connected (Note that wireless OSP does not use third-party for call delivery to the NGCS, cellular, or mobile, including PCS, or national, regional, and international providers, resellers.)	AT&T Mobility, Verizon Wireless (Note that in some areas Sprint is directly connected to the NCGS and others; they use a third-party aggregator. There is no universal assumption for direct and indirect OSPs.)	Voice, text, data, telematics, and IoT; video, not addressed in the first edition of this book.	Standard text* options vary with the 9-1-1 SSP and NGCS interfaces; pANI NPA 555-XXXX arrives at PSAP before actual location from third-party location provider via e2 datalink.
Wireless—third-party connected (Note that third-party engineers call delivery including signaling option. Cellular or mobile including PCS. National regional and international providers, resellers.)	Sprint PCS, T-Mobile	Voice, text, data, telematics, IoT; video not included in the first edition of this book.	Standard text options vary with the 9-1-1 SSP for NGCS interfaces. pANI NPA-555-XXXX arrives at PSAP before location from third-party location provider e2 datalink.

*The most common text-to-9-1-1 solution deployed has been SMS-based 9-1-1. This is an evolving topic.

One type of wireless connection is direct from the wireless OSP to the NG9-1-1 network. When a NG 9-1-1 network is engineered with sufficient capacity to accept calls across a region, state, or multiple states, there is incentive for the wireless OSP to connect to two diverse NG9-1-1 data centers or COs.

Other wireless OSPs choose to use third-party aggregators. The aggregator sends the calls to the NG9-1-1 network. Wireless OSPs can deploy services more efficiently over larger diverse facility paths to reach many more PSAPs. Wireless OSPs can directly access NG9-1-1 NGCS core technology with bandwidth designed for call volume and types of services to be completed. Facility and trunk engineering is an important topic of the negotiations between wireline OSPs and 9-1-1 SSPs. Wireless OSPs in conjunction with 9-1-1 SSPs should engineer their trunk groups to keep one incident from overloading the facilities for all callers over the NG9-1-1 9-1-1 SSP's footprint.

Wireless location accuracy is a critical concern. All wireless OSPs use MPCs to provide caller location. Calls initially arrive at the PSAPs with pANI numbers. This is what is known as phase I wireless to the PSAPs. The MPC supplies the dialed number of each caller along with the location (X, Y coordinates for the call to be plotted on a map via e2 datalinks). The number of the caller is important in the event a connection is dropped and the PSAP call taker needs to get back in touch with the caller. This is known as phase II wireless. Chapter 10 discusses mandatory FCC required improvements in emergency location services. Many groups, including academic laboratories facilitating cooperation between wireless OSPs and universities, are researching additional options.

Third-party aggregators may choose a trunking and facility aggregator to transport traffic to deliver the wireless calls. Level 3 is one such OSP, but there are additional providers offering aggregation service. Getting all parties together to make decisions about engineering, signaling, location services, and testing, including connection agreements, adds layers of complexity and time to negotiations. There are technical considerations related to signaling types in the third-party scenarios; these are addressed in Section 5.3.

5.1.3 VoIP OSPs

VoIP OSPs use IP-based networks and applications to provide service. Such OSPs offer their customers the opportunity to replace traditional circuit-switched telephone service with a service carried over a packet-switched IP network. When such services were introduced in the early 2000s, the providers included a caveat in their marketing brochures stipulating that 9-1-1 emergency services were not a part of their offerings. The FCC intervened and made it mandatory for VoIP OSPs to add 9-1-1 to their service. This resulted in modifications to the E9-1-1 wireless call delivery model NENA developed to adapt to VoIP technology. Chapter 1 provides further insight into VoIP technology and its implications for emergency services. This chapter aims to identify two types of VoIP OSPs that connect to NG9-1-1 networks.

Most VoIP OSPs use third-party providers to facilitate call delivery and location services. The dominant third-party VoIP emergency service providers in the United States are West Public Safety, Comtech TCS, and Bandwidth.com. Each provider has a confidential list of VoIP OSP clients. Confidentiality creates planning, engineering, and operations challenges for 9-1-1 SSPs. Each VoIP OSP needs to test from their OSP customers' lines and locations end-to-end to ensure accurate call completion along with the correct location. There are two types of VoIP OSP: fixed-VoIP providers and mobile-VoIP providers (MVP).

The category of fixed VoIP is significant, because callers often are not overtly aware that their telephone services are using VoIP technology. When consumers choose bundled telephone, Internet, and Wi-Fi services in homes and offices, the former traditional phone lines may be incorporated or bundled into that service offering. A home or office phone can be connected to the same jack as it was the day before the transition to VoIP services, using the outside plant connecting the calls to a traditional ILEC, RLEC, or CLEC switch, yet the VoIP providers no longer carry the calls directly to a traditional CO. While customers are drawn to VoIP because the cost is lower than traditional wireline service, the level of service may change. A storm that knocks out commercial power did not affect the 48V telco connection that would stay active to a home phone line. VoIP service no longer has access to that same power and is likely to fail; likewise, uninterrupted power to a remote VoIP Hub, may last two to four hours but not indefinitely. Companies such as AT&T Uverse, Verizon FIOS, and Comcast Xfinity are using fixed VoIP as part of their offerings. While a home or business location has not changed, the method of getting the location information to the NG9-1-1 database may have changed in a major way. Location accuracy is one reason some customers retain wireline service on ILEC or RLEC traditional networks. There are major fixed-VoIP providers for businesses, including AT&T Collaborate VoIP and Verizon Business VoIP, which may use their own facilities and provide location updates as they do in the traditional ALI database.

The second type of VoIP service allows for portable phones/devices that customers can take anywhere in the world. The customer is responsible for registering their location with the VoIP OSP for location accuracy in real time. Each OSP needs to be identified and agreements made before the services can be transitioned. Many large enterprises are transitioning to VoIP services. Refer to Chapter 6 for enterprise routing and location requirements. State commissions have or exercise little jurisdiction over VoIP providers in general. FCC regulations are important to the quality of emergency services for VoIP customers. See Table 5.3.

5.1.4 OSP Services

IoT access technology allows emergency access to the ESInet from the OSP that provides transport. These calls route through wireless OSPs. Table 5.4 identifies these services and how they are handled by different types of OSPs. Telematics testing started in the United States as part of a Federal Department of Transportation

Table 5.3 VoIP OSPs*

Type of OSP: VoIP

Category	*Example*	*Service*	*Attributes*
VoIP-fixed: This includes "fixed" in homes and businesses that are sometimes bundled: television, Internet, and Wi-Fi access.	AT&T Uverse, Comcast Xfinity, and dozens of fixed-VoIP providers Enterprise customers using VoIP 9-1-1 systems routed directly to the NG9-1-1 networks are addressed in Chapter 6.	The VoIP provider assumes responsibility to provide location information to the 9-1-1 SSP database, often through third-party location providers.	Third-party location provider; pANI NPA-222-XXXX arrives at PSAP before actual location from third-party location provider over e2 datalink.
VoIP-Portable VoIP used by the OSP. Portable phones operate like wireless devices. The VoIP provider assumes responsibility for providing location information.	AT&T, CenturyLink, Verizon, Vonage, and hundreds of individual VoIP providers serving individuals and businesses with VoIP PBXs with phones used at desktops around the world.	Voice services for emergency services; location is mandatory (caller must register location into the OSP database for proper location to arrive so that the 9-1-1 SSP can route to the appropriate PSAP).	Third-party location provider; pANI NPA-222-XXXX arrives at PSAP before actual location from third-party location provider over e2 datalink.

*https://en.wikipedia.org/wiki/Enhanced_9-1-1#The_911_Act, "In recent years there have been numerous important developments in E911 solutions for IP phone technology. The more noteworthy of these developments include on-site appliances that automate and simplify E911 management for enterprise IP-PBX systems, reducing administration, ensuring that IP phone locations are always up to date, thus helping enterprises meet their E911 obligations; IP phone tracking that automatically assigns locations to IP hard phones, soft phones, and wireless phones as they move on the corporate network using layer 2, layer 3, or wireless LAN discovery. Support for remote employees, allowing off-campus users and teleworkers to update their locations in real time directly from their IP phones; support for phone mobility, to ensure accurate E911 services for employees that move IP phones between locations, share line appearances between multiple devices, and log into IP phones on the fly; security desk routing and notification functionalities that deliver 911 calls and custom email alerts to on-site security personnel, notifying them of the emergency and providing them with the caller's precise location information; advanced E911 call management and reporting features, such as misdial protection and call recording, to improve solution performance and administration. VoIP and 911 issues are also relevant to telecom relay services utilized by individuals with disabilities.

Table 5.4 Significant Services provided by OSPs

Type of Services: Additional Wireless Services Tables 5.1–5.3			
Category	*Description*	*Service*	*Attributes*
Telematics	Vehicles with emergency access; manufacturers have varied implementations nationally and internationally;*	Initially emergency access was to a manufacturer's call center such as General Motors On-Star. Advanced access is available through OSPs to the appropriate PSAP using access and location technology selected by the manufacturer(s).	Standards vary with the 9-1-1 SSP; third-party location arrives over e2 datalink.
IoT	Various products allow for persons or devices to interact with a smart device requesting emergency services. Protocols vary; products include smart phones with voice integration such as Siri, Alexa, Cortana, and Google Assistant.**	IoT is consumer- and technology-driven; the products are not necessarily designed with emergency services as part of the technology.	Standards vary with the 9-1-1 SSP. Third-party location arrives over e2 datalink.

*Oorni, R., and A. Goulart, "In-Vehicle Emergency Call Services: eCall and Beyond," IEEE Communications Magazine, New York, NY, Jan. 2017, Vol. 55, No.1, pp.159–165.

**Kudievskiy, A., "Three Reasons Every Business Should Consider Voice Technology," Forbes Community Voice, Aug. 14, 2018, 7:30 a.m., Forbes Media, L.L.C. Jersey City, NJ, 2018.

(DOT) NG9-1-1 project[6]. The overarching goal of the telematics project was to save lives on the highways. The wireless OSPs that handle the traffic may be directly connected to ESInets or may be sent the requests for emergency assistance through a wireless OSP that uses a third-party provider or aggregator to reach the NG9-1-1 network. Portable and fixed devices are being integrated into NG9-1-1 testing scenarios to proactively get help in medical emergencies and to aid in locating people calling for emergency support in large buildings. Smart watches are sending emergency services calls through the E9-1-1 and NG9-1-1 networks to PSAPs. Chapter 10 discusses the emerging technologies.

5.1.5 OSPs Serving an Area

9-1-1 SSPs need to know which OSPs are serving their areas of responsibility. NENA has an online database document for self-identification of all OSPs that provide 9-1-1 service. It includes contact information and a description of the type of OSP or carrier by name. This is a necessary reference, yet individual entries are missing

6. https://www.its.dot.gov/research_archives/ng911/ "Research Overview-The nation's current 9-1-1 system is designed around outdated telephone technology and cannot handle the text, data, images, and video that are common in personal communications and critical to future safety and mobility advances. In addition, 9-1-1 call centers cannot transfer calls from one call center to another when call volume exceeds the available resources. The nation's 9-1-1 system is in need of a significant overhaul. By capitalizing on recent technology advances, the U.S. Department of Transportation (DOT) has produced a design and transition plan for a next-generation 9-1-1 (NG9-1-1) system. The NG9-1-1 Initiative has established the foundation for public emergency communications services in a digital, Internet-based society."

or are not always up-to-date.[7] The purpose of the NENA Company ID (CID) Registration Service is as follows: CIDs allow 9-1-1 centers to request emergency network services, such as busy-line interrupt and call trace, when required to save lives and protect property. Additionally, CIDs help to ensure that accurate caller location information can be provided to 9-1-1 centers (PSAPs) and first responders when a user calls 9-1-1. CIDs also facilitate timely restoration of service during certain network or equipment outages. Listing the registered CIDs in the NENA Company ID database allows access infrastructure providers and data providers, states, counties, cities, and PSAPs to access and use the Company ID information. The NENA Company ID in the 9-1-1 ALI record allows the PSAP to quickly identify the access infrastructure provider and/or data provider for the caller's telephone number and to determine the 24x7 number of the company for emergency contact needs. The key fields include: company ID number, company name, type of carrier, status, and 24-by-7 contact information. the reference is self-monitored by the OSPs, and the contact information or company name is often out-of-date, but it is a good place to start researching.

State agencies, PSAPs, and the LERG[8] are resources to identify the OSPs that serve an area. Each state has a commission that regulates utilities, including OSPs. Many states have a specific staff or bureau for 9-1-1. All states have legal certification processes for OSPs to operate within the state. The names and contacts for OSPs operating within a state are generally known because of tariff filings, interconnection agreements, and/or billing and payments related to 9-1-1 surcharges. The contact information can be outdated, and the state commissions are not always current on acquisitions or business failures. Great care must be taken in identifying the OSPs and a combination of all these sources is necessary.

The PSAPs may provide information about the OSPs whose customers they serve. PSAPs generally know each OSP they serve even if the OSPs do not route calls to them directly. Minimally, the PSAPs and their managers know the dominant OSPs in their area of responsibility. The PSAP may receive direct OSP payment of the 9-1-1 fees, but that arrangement is generally made through a billing agency and does not easily lead to the contact that OSPs must connect to a NG9-1-1 network.

When the area that a new 9-1-1 SSP plans to serve has an existing 9-1-1 SSP, that entity often files an annual report with the state and the PSAP. Filings include diagrams, although the details vary significantly.

Every connected OSP should have an access service request (ASR) on file with the existing 9-1-1 SSP's marketing and/or engineering departments to be connected

7. https://companyid.nena.org/page/AboutCID, "NENA Company Identifiers (also known as Company IDs or CIDs) allow 9-1-1 centers to quickly identify the telephone company or access infrastructure provider responsible for a particular telephone number. The evolution of telecommunications has necessitated the need to display Company Identifications (CIDs) of the associated Access Infrastructure Providers and Data Providers for each telephone number to the Public Safety Answering Point (PSAP). This need is driven by two factors: Speed of Identification by PSAPs, Use by ALI DBMS Providers. A national NENA Company ID registration system was implemented by NENA Standards work groups in 1996, and subsequent work has made this service a part of this website for general access. This approach is intended to support standardization of NENA Company Identifiers (CIDs), and to supply a single point of administration for the Company ID file content and update."

8. LERG is a Trademark of iconectiv.

to any SR or directly to a PSAP. An OSP may use the NENA PSAP ID reference[9] list to notify the PSAP when it plans to start doing business in its jurisdiction. The OSP uses the PSAP name and ID tables maintained by NENA unless it has information from another source. Resellers are often on named lists although their calls route from their serving OSP.

The LERG has the names of most major ILECs, RLECs, and CLECS and some wireless OSPs. A subscription to the LERG through iconectiv is essential for anyone planning to deploy a NG9-1-1 network. Additional sources are available. Figures 5.1–5.8 use information from these sources.

Critical data includes, but is not limited to, the following:

- Identification of each OSP;
- Identification by area code and NPA NXX of most wireline OSPs;
- Identification of each originating carrier number (OCN) or OSP identification number;
- Identification of each (CLLI™[10]) for each OSP's switch;
- Identification of the type of OSP switch per CLLI;
- Identification of and CLLI for each SR;
- Host remote switching relationships;
- Identification of split exchanges at PSAP boundaries;
- Identification of SRs with SS7 PCs).

All references are critical; this list is not all inclusive. In places where information is in conflict, firsthand research must be done to complete plans. Below is a supplemental list of resources to locate OSPs that serve and area.

1. Area codes: http://www.area-codes.com/search.asp?frmNPA=618&frmNXX=294&frmCity=&frmState=IL&frmZip=&frmCounty=&frmCompany=&search.x=0&search.y=0;
2. U.S. telco database: https://www.telcodata.us/search-area-code-exchange-by-company-state?company=odin+telephone&state=IL;
3. Local calling guide: https://localcallingguide.com/lca_prefix.php?npa=630&nxx=941&x=&ocn=®ion=&lata=&switch=&pastdays=&nextdays=.

Each OSP in a serving area must be migrated from the existing network to the new NG9-1-1 network. No customer requiring emergency services can be left behind. There are commission exchange boundary maps. An exchange is a term used by the legacy telecommunications carrier or OSP boundaries to identify the call-originating switch in a CO. The maps are essential to identifying OSP boundaries within States, county boundaries, or PSAP boundaries and to identifying potential

9. https://www.nena.org/page/PSAP_Registry, "The NENA PSAP Registry, developed in conjunction with DDTI, is the overall name for a NENA database that contains information on PSAPs throughout the United States, and may expand to Canada in the future. This central database will continue to fulfill the NPR's original purpose of supporting PSAP personnel to locate and contact other PSAPs during critical call transfer situations."
10. CLLI is a trademark of iconectiv.

split exchanges. National NENA has state and regional chapters with members who regularly meet to plan for changes, provide certification and education, and work with local, state, and federal officials. For example, in Illinois there is the Illinois Public Safety Telecommunications Association (IPSTA). The state is divided into regions with a lead contact for each region. The names and numbers of leaders are posted; quarterly and annual meetings are planned; and members share information. Emergency communications professionals volunteer to write NENA standards, develop processes, and maintain the emergency services networks. Join NENA, at NENA.org or EENA, at eena.org, to access references and standards online. NENA members are in contact with similar organizations worldwide. The bottom line is that *a considerable number of hours are needed to locate the OSPs in the 9-1-1 SSP footprint.*

OSPs are an integral part of the emergency communications service. There is no way to plan for NG9-1-1 without working with each OSP

Sections 5.2–5.12 addresses the most common issues to be addressed between OSPs and 9-1-1 SSPs at the onset of any project.

5.2 Facilities

One of the first building blocks for NG9-1-1 is the ESInet, which is described in Chapter 3. OSPs need a description and location and the identification of data centers or COs connected to the ESInet that house the NGCS for the NG9-1-1 network. The addresses and CLLI for the NGCS and the means of connection are equally important. The 9-1-1 SSP needs to specify the POIs along the ESInet where OSPs can connect. The POIs must be secure, powered, grounded, environmentally supportive, and ready to connect OSP equipment and provide switches and or routers to allow the calls to flow in any direction around the ESInet should a path along the ESInet fail for any reason. NG9-1-1 standards for connecting facilities and using switches and routers allow for nearly every transport technology (e.g., fiber, copper, wireless, microwave, and satellite). Internet access is an option although nearly all 9-1-1 SSPs require secure, dedicated access. Redundancy is also a requirement regardless of the chosen transport technology. The OSPs need to know that if one of the data centers, POIs, or interfaces fails, the design allows their calls to automatically reroute to complete. Most NG9-1-1 NGCS equipment vendors and 9-1-1 SSPs will not (or should not) accept connections below a T1 or E1 level. Many Ethernet providers do not sell access below 10 megabits of bandwidth over an IP backbone network. Ideally the ESInet is built on a fiber ring; that ring should not be a collapsed ring.[11] Each OSP decides how and where it chooses to terminate its facilities. Each OSP determines how much access bandwidth it requires. The NG9-1-1 network accepts the emergency calls from each OSP and routes them to the correct PSAPs. ESInets with duplicated NGCS benefit OSPs that currently have to complete calls to many legacy SRs today without SR redundancy. OSPs need to understand they can connect at a regional, state, or multi-state level. Many wireless OSPs are

11. A collapsed or folded ring refers to a ring architecture where the physical fibers used for working and protection both reside in the same physical bundle or buffer tube. The result of this is that typically in the event of a fiber cut, the entire ring will be taken down.

connected to a whole state using only two POIs in this manner. Such connections are technically efficient and cost-effective. Once transport bandwidth is increased for access, a larger number of calls can route into the NGCS more efficiently. To properly size the ESInet and transport to the PSAPs, it is essential to know how much traffic is forecast to originate from any particular OSP in the busiest hour of the busiest season projected for 9-1-1 services. The 9-1-1 trunk group busy hour is typically not the same as most other trunk groups served by the same switch. The 9-1-1 SSP BCF is set at an appropriate number of queries per second to allow good calls to go through while preventing an overload to any NGCS component and/or terminating PSAP. The number of simultaneous calls and the average time per call dictate the minimal traffic to let through the BCF. State rules vary on whether to allow emergency calls to remain in queue to a PSAP or allow the caller to hear a regular busy (60 interruptions per minute), fast busy (120 interruptions per minute), or announcement so they know to redial. Testing the BCF is an important function before adding live traffic to any NG9-1-1 network. There is no way for the PSAP or a new 9-1-1 SSP to know if there is pent-up demand that could be released when the new network design is implemented. Networks have become overloaded when unlimited wireless demand has full access to limited positions per PSAP. As the network is cut over for new services, the 9-1-1 SSP should consider adding services incrementally and monitoring closely for impact. If a PSAP already handles text to 9-1-1 and voice today, cutting them together is expected. When using T1s or E1s over copper, a full T1 or E1should be ordered. If the transport connection is via Ethernet, no less than 10 megabits of capacity are usually available for purchase. Each OSP's ASR must include the type of transport and the desired bandwidth. Finally, it is the OSP's option to choose access facility diversity.

5.3 Signaling

OSP-originating signaling options for emergency services include SIP, ISDN-PRI, SS7, and MF. SIP signaling interfaces connect directly to the BCF before passing into the NGCS for completion. All legacy signaling interfaces, ISDN-PRI, SS7 and MF, must connect to the LNG for signaling conversion to SIP before passing into the BCF then the NGCS for completion. Each OSP ASR must specify the type of signaling to be used for each originating OSP to connect.

5.3.1 SIP

The NG9-1-1 standards use SIP signaling within the NGCS core network extended to i3PSAPs. In the best case, the OSP chooses SIP signaling to deliver calls to the core. SIP enables critical calling-party information including location to be carried directly to the i3PSAPs, whereas other signaling methods require that such information comes to the PSAP via separate paths. SIP also reduces the need for voice-to-voice communication at critical handoffs to dispatch agencies and between i3PSAPs. However, most wireline OSPs do not have SIP-capable switches unless they are using Softswitch technology to originate calls or unless they purchased SIP interfaces for their circuit switches to use with IP PBX customers. OSPs using SIP signaling must connect directly to the BCF for security reasons. Figure 5.1

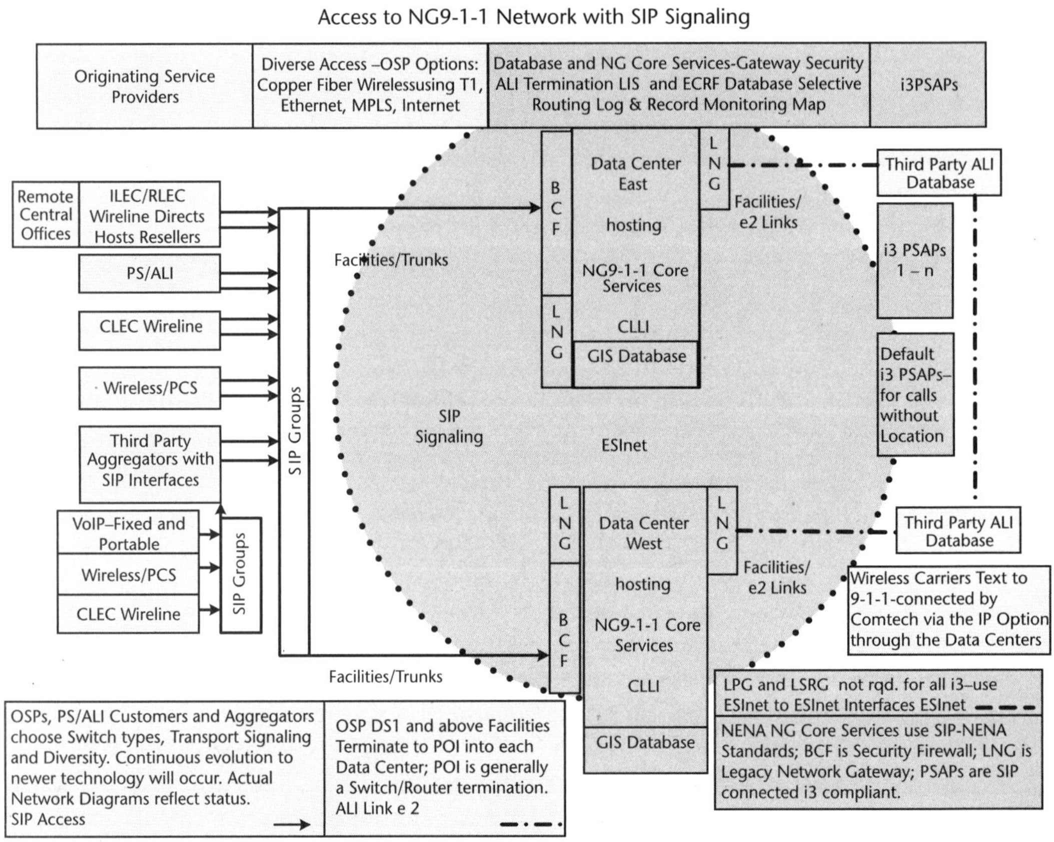

Figure 5.1 OSPs using SIP signaling to the NG9-1-1 network.

illustrates the signaling path for SIP calls, showing various types of OSPs sending their 9-1-1 calls as IP-based SIP calls that are then carried over a SIP path into the BCF. Today, calls do not include location information in the body of the SIP messages, and, based upon the identification of the facilities that bring in the call, the NGCS chooses a PSAP and routes the call to that PSAP. At the same time, a dip of an onsite NGCS ALI database for wireline calls or a third-party ALI database provides the location of the wireless calling party, and this location travels around the fiber ring to the answering PSAP to be displayed on the screen of the call taker. The e2 datalink signaling interface is always through the LNG.

5.3.2 ISDN-PRI

ISDN-PRI is the second preferred method of signaling. Many wireline OSPs have ISDN-PRI trunking available in their switches. ISDN-PRI provides faster call setup than the older MF signaling technology and does not require the out-of-band SS7 signaling links and signaling transfer points (STPs) required for SS7. Figure 5.2 illustrates the facilities and trunks for calls entering the NGCS from an OSP using ISDN-PRI trunks. In Figure 5.2, calls arrive at the ISDN-PRI trunk group from an OSP. The ISDN-PRI trunks terminate on the LNG, and the calls that they carry are translated into IP calls by the LNG and handed off to the BCF. From the BCF, the call is processed in the NGCS and routed over the fiber ring (or other transport

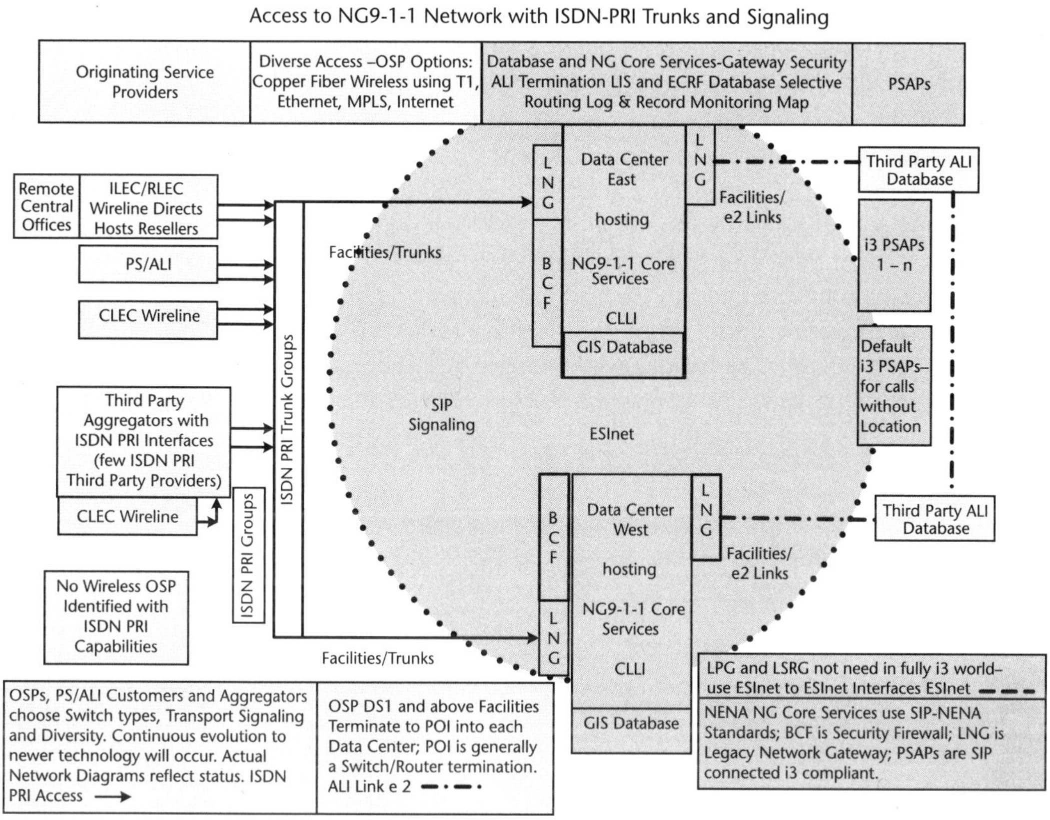

Figure 5.2 OSPs using ISDN-PRI trunks and signaling to NG9-1-1 network.

technology) to a PSAP designated by the NGCS using information associated with the ISDN-PRI trunk facilities. While the call is being delivered, the ALI database via e2 datalink is dipped to learn the location of the caller. This location information is then routed to the i3PSAP over the fiber ring and displayed on the screen of the call taker.

5.3.3 SS7

SS7 is the predominant signaling type in use for OSP 9-1-1 access in 2020. SS7[12] PCs must be assigned per OSP switch and shared between the OSP and the 9-1-1 SSP for calls originating and terminating to each NGCS. The 9-1-1 SSP must either be an SS7 service provider with STPs, or legally contract with a third-party SS7 provider for SS7 services. The physical addresses of the STP pair must be provided along with CLLI and PCs to each OSP using SS7 signaling. Existing SS7 A-links for

12. PC assignments are available for purchase from iconectiv. The LERG includes the 9-1-1 PCs and associated CLLI for any OSP requiring connections to the NGCS using SS7. The originating PC (OPC) must be provided by each OSP for each SS7 trunk group using SS7. If the OSP is a wireline carrier with host remote COs, the PC for the host is the one required to be identified. Each remote CO uses the same PCs and A-Links as the remote. The A-links only connect the host. The 9-1-1 SSP applies to iconectiv and pays to obtain its PC assignment for the NGCS. The NGCS uses SS7 A-links from each STP into each NGCS through the LNGs. Figures 5.3, 5.7, and 5.8 show the connections.

the OSP's switches are used. New SS7 A-links are established between the 9-1-1 SSP's STP pair to each NGCS at the LNG interface. 9-1-1 SSPs can use dedicated diverse SS7 A-links or SIGTRAN[13] virtual links between the STP pair and each NGCS. Before planning and contracting for SS7 connectivity, some NGCS vendors do not provide SIGTRAN interfaces with their LNG products. Dedicated facilities and A-links require transport diversity to each NGCG. Figure 5.3 illustrates the signaling path for a call whose OSP uses SS7 signaling. In Figure 5.3, a call originates from one of the OSPs and is translated into an SS7 call for transport into the LNG with signaling channels on the SS7 links as well as bearer channels from the trunk both terminating on the LNG. The LNG translates signaling and media to SIP and RTP respectively and hands the call to the BCF. The call is sent to the PSAP chosen by the NGCS based on identifiers associated with the trunk facilities, and simulta-

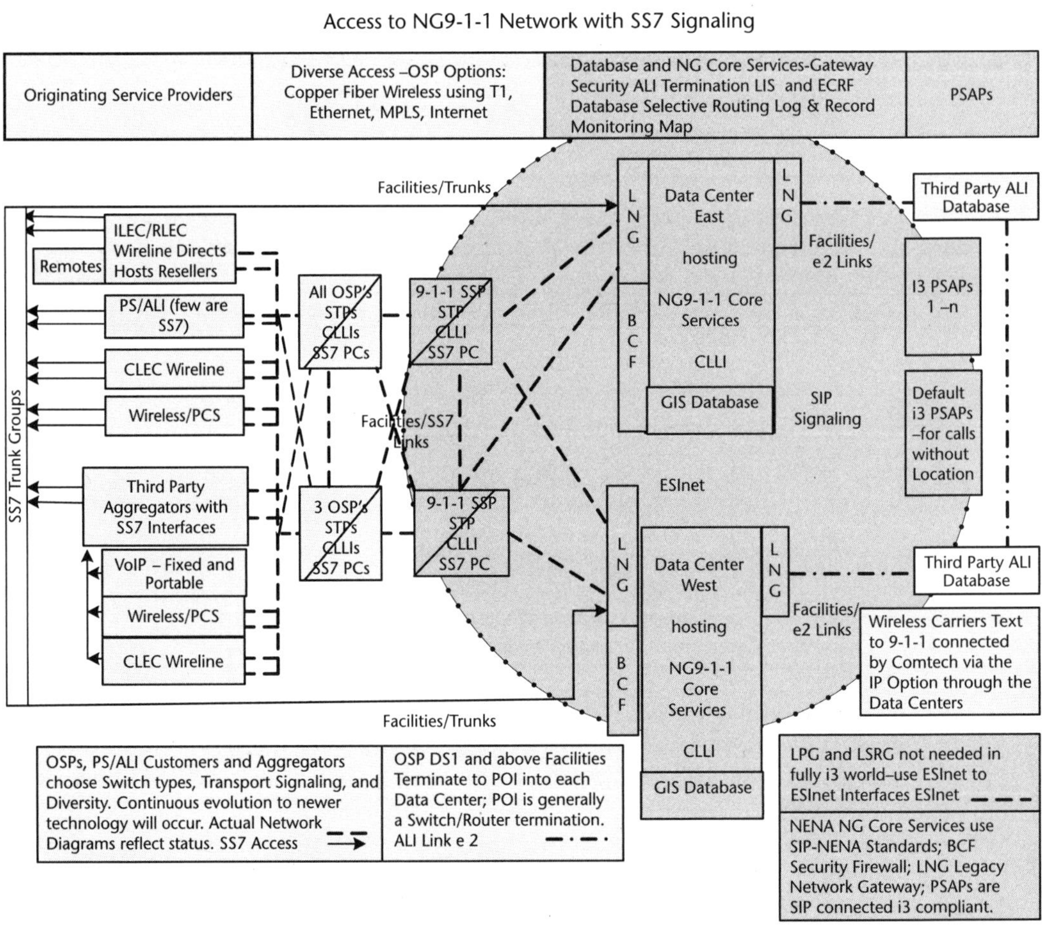

Figure 5.3 OSPs using SS7 trunks and signaling to NG9-1-1 network.

---

13. The IETF has defined the SIGTRAN protocol suite that implements levels 2, 3, and 4 protocols compatible with SS7. Sometimes also called *pseudo SS7*, it is layered on the stream control transmission protocol (SCTP) transport mechanism. SIGTRAN is not generally the best option for 9-1-1. Note: In some cases individual NGCS vendors do not support Sigtran integration for their NG9-1-1 solutions.

neously the third-party ALI database is dipped via e2 datalink to learn the location of the caller that is sent to the PSAP via the ESInet.

5.3.4 MF

At one time MF signaling was considered the optimal standard for access by OSPs into the 9-1-1 and E9-1-1 network. At the inception of 9-1-1, MF was the newest most reliable access signaling technology available. The first use of SS7 access technology for 9-1-1 was in 1996 and 1997 in the Chicago, Illinois, E9-1-1 network architecture. By that time, SS7 reliability had improved and the Chicago network was designed with SS7 and diverse access from each OSP originating office to dual legacy SRs in the architecture. The introduction of SS7 into originating offices meant the interoffice trunking and call setup was faster than could be done with MF signaling and trunking combined. Customers calling 9-1-1 may perceive any slight delay in hearing the ringing tone to mean that the call failed, and they tend to hang up and/or regenerate a new call rather than wait for connection, thus risking lives. So, unless a whole CO is using MF signaling, it is not recommended to leave 9-1-1 on MF signaling and have all other calls using SS7 or more modern SIP signaling for call setup. The standards allow for MF signaling to be deployed. It is the choice of the OSP. The ISDN-PRI call flow in Figure 5.3 is similar to MF signaling, which interfaces to the LNG to reach the BCF and NGCS.

5.3.5 CAMA

When a legacy SR remains in the access network, the standards allow the connection from the legacy SR to use CAMA signaling to the LNG of the NGCS. The OSP can decide to continue to use the CAMA connection or can convert the legacy SR to NG9-1-1 interface to an ISDN-PRI to NG9-1-1 trunk group or an SS7 to NG9-1-1 trunk group for greater efficiency and safety. CAMA trunks are an older component of the 9-1-1 and E9-1-1 architectures. CAMA trunks have been known to fail and cause PSAPs to completely lose connectivity. CAMA test gear is manufacturer-discontinued. If CAMA trunks continue to be used, it is essential to test calls through that technology. The testing should include a SLA requirement for the OSP to do monthly routine testing and verify that everything is operational. CAMA trunks terminate to the LNG before routing to the BCF, the NGCS, and then the PSAPs. See Figure 5.4 for CAMA signaling routing examples.

5.3.6 e2 Datalink Signaling

Location services are a critical part of E9-1-1 and NG9-1-1. In the future, the caller's location will be derived from the originating device. At the present time, each type of OSP and 9-1-1 SSP combination has unique ways to display the caller's location to the PSAP in near real time, using duplicated diverse e2 datalink ALI links to each third-party location provider. This section describes how wireless, VoIP, and some CLEC OSPs use the services of third-party providers to derive the location of callers and in particular how these third-party providers are connected to the NG9-1-1 LNGs.

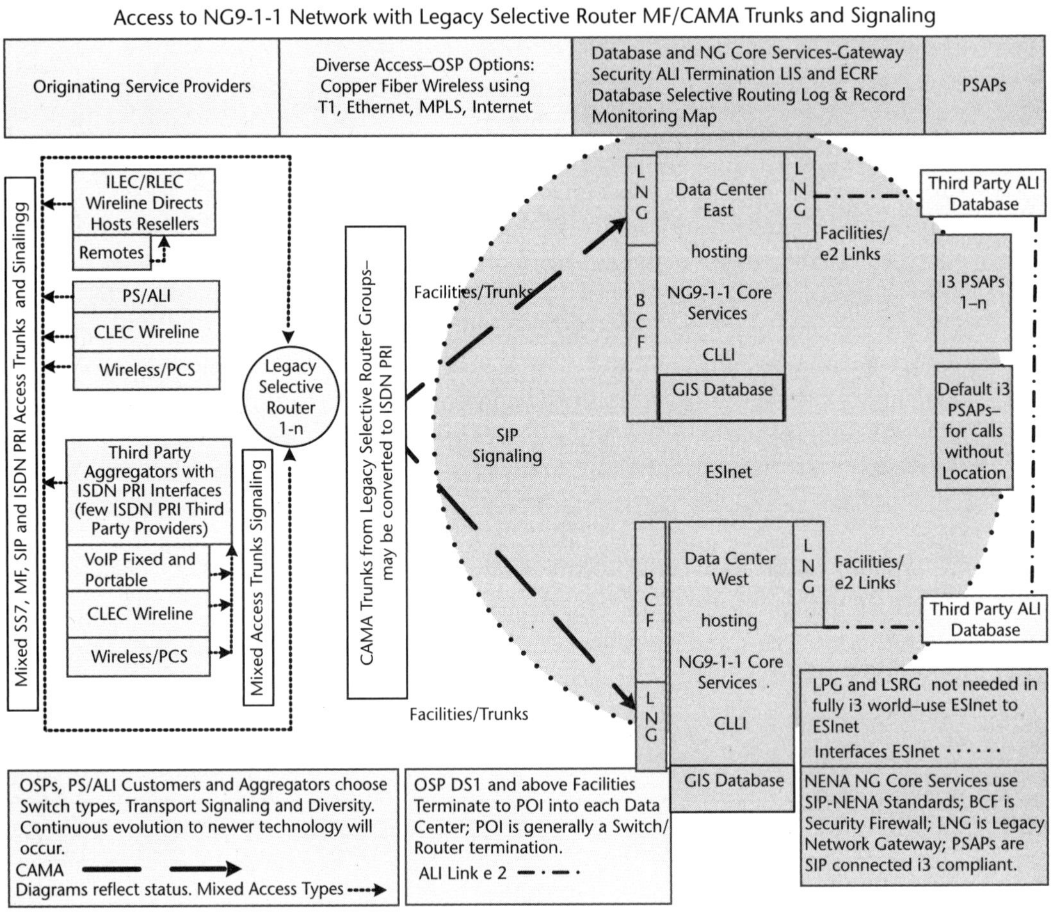

Figure 5.4 OSPs using legacy SR CAMA trunks and signaling to the NG9-1-1 network.

When planning for the migration of a specific geographic area to NG9-1-1 begins, the OSPs and their third-party providers must be notified. Start with the major OSPs in the footprint and learn the name of the provider each uses to deliver ALI service.

Wireless OSPs all use third-party location providers. Wireless callers may be roaming anywhere in the world. This means that the phone number associated with the wireless devices cannot be used to locate the caller, nor can it be used to identify the PSAP that needs to handle the call. The X, Y coordinates of the cellular tower from which the call arrives, or the X, Y coordinates of the caller's handset, as determined by the location provider/carrier, identifies which PSAP should receive the call. The PSAPs need to know the caller's actual phone number to call back the originator should the connection be lost. Wireless OSPs are usually provided with the pseudo NXX code of 555. The line number ranges go from 0000 to 9999. A series of numbers are assigned to each PSAP. The e2 datalink is queried by either a legacy PSAP or the NGCS for NG9-1-1 to request the actual phone number of the caller. If there is a legacy PSAP, the LPG is the interface to the PSAP from the NGCS, and the PSAP continues to directly query the location services as they have since the inception of E9-1-1 wireless phase I.

Since the caller may be moving, the PSAP that initially receives the call may need to transfer the call to a different PSAP that can best perform the dispatch function. There are specialty companies that manage third-party database data entry for large wireless or VoIP customers. We do not detail those options here.

VoIP OSPs users can operate in a fashion similar to that of the wireless OSPs. The third-party database sometimes relies on location to be registered by the portable VoIP customer that owns the handset or the agent for the VoIP customers. For fixed-VoIP customers, the OSP ensures that the addresses are preloaded into the VoIP third-party provider databases they choose. VoIP third-party providers use pseudo NXX codes starting with 222. The 0000-9999 range of the line numbers are assigned to the OSP's VoIP third-party provider to share with the 9-1-1 SSP to direct calls to the correct PSAP. The accuracy of the incoming location can depend on the accuracy of the X, Y coordinates of the caller's location. The name of the third-party's VoIP location center is the VPC. The FCC is expanding location services requirements to require Z-level location

CLECs have found that outsourcing their customer address locations to third-party providers for inclusion in the ALI database can relieve them of their direct customer location responsibilities. The location accuracy for VoIP has proven to be less reliable than the method used by ILEC, RLEC, and other wireline CLEC OSPs in general. Less reliable here means that calls may arrive at an adjacent or wrong PSAP and need to be transferred. CLECS often serve major businesses with multiple locations and multiple buildings, where the update process can be delayed. Any delays in database entry and validation may cause calls to arrive and no location to arrive with them, known as no record found (NRF). Figures 5.1–5.8 show and refer to default PSAPs that must be designated and designed to handle the NRF callers.

The 9-1-1 SSP with PSAPs over large geographies is responsible for assigning a primary default PSAP and alternate default PSAP whose purpose is to receive calls when no location or PSAP identification arrives with the call. The default PSAPs receive all calls without location, and these PSAPS are trained to investigate voice and text calls to determine each caller's location and to forward the calls to the proper PSAP for dispatch using a warm handoff. The default PSAPs can generate a trouble ticket for each incoming call with no location so that the database administrator and/or technical support person can determine if duplicated e2 datalinks are down, or the caller's SO-driven address location has not arrived at the NG9-1-1 database. Chapter 10 addresses advances in location accuracy. Figure 5.5 illustrates the e2 datalink signaling and location methodologies available, including default PSAPs. The databases of the third-party providers, including the VPC and the MPC, are all shown on the left-hand side of Figure 5.5, although these databases have been shown in the previous diagrams (Figures 5.1–5.4) on the right. This was done simply to make Figure 5.5 easier to read, and not to suggest any architectural differences.

5.4 Trunking

The 9-1-1 SSP needs to know the number of trunks each 9-1-1 OSP provides into the NG9-1-1 network. Trunking and routing information needs to be specifically

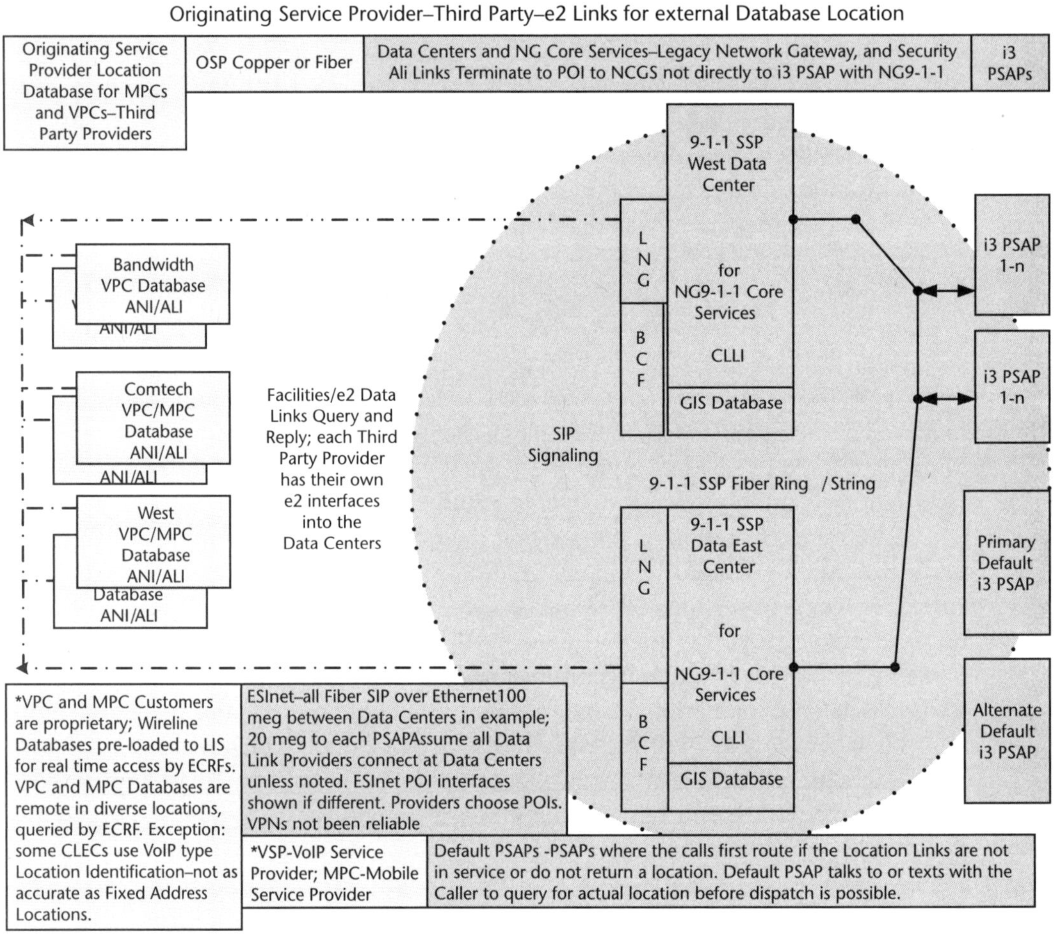

Figure 5.5 VPC MPC e2–datalinks for PSAP to query for location callers without fixed addresses in NGCS.

identified by the OSP at the time of ASR submittal if not before. In general, the OSPs design the trunk group sizes based on the anticipated emergency services traffic. Most state commissions and the FCC require that no more than one emergency call in 100 is blocked in the busiest hour of the busiest rolling 20-day busy season of the year. This is called P.01 grade of service (GoS). The P comes from Poisson[14] traffic engineering theory. NENA standards suggest that no less than two trunks be provided for access by any OSP for local access, wireline, or wireless switch into the

14. Telecommunications traffic engineering, teletraffic engineering, or traffic engineering is the application of traffic engineering theory to telecommunications. Teletraffic engineers use their knowledge of statistics including queuing theory, the nature of traffic, their practical models, and their measurements and simulations to make predictions and to plan telecommunication networks such as a telephone network or the Internet. These tools and knowledge help provide reliable service at lower cost.

The field was created by the work of A. K. Erlang for circuit-switched networks but is applicable to packet-switched networks, as they both exhibit Markovian properties and can hence be modeled by, for example, a Poisson arrival process.

The crucial observation in traffic engineering is that in large systems the law of large numbers can be used to make the aggregate properties of a system over a long period of time much more predictable than the behavior of individual parts of the system.

9-1-1 network. Traffic demand using engineering formulae determines the number of trunks required to meet the required GoS. The 9-1-1 SSP cannot always see the traffic engineering data from the access side of the network and thus generally accepts the number of trunks required from each OSP on the ASR. If negotiations are required, that should be done in the planning phase of the project. In E9-1-1 networks there is seldom more than one legacy SR accepting trunk groups from any single OSP switch. In NG9-1-1 networks the number of equivalent SRs, part of the NGCS, is typically two, and they generally cover a much larger area. The concept is that if one NGCS, a component of the NGCS, or the complete data center fails, the mate can assume the full traffic load from the first failing NGCS or component. Typically, following a logical model developed for SS7 networks, each NGCS should be engineering to carry the full load anticipated by all OSPs demanding service. An NGCS is considered loaded at 40% so that if the first routed NGCS fails, the mate's 40% load can be added on top and carried immediately. This brings the load to 80%. The remaining 20% capacity is set aside to handle any surge or peaks in demand from a major unforeseen event such as a hurricane, earthquake, or manmade or natural disaster. The bandwidth between data centers is another calculation to ensure that all the traffic from all OSPs and to all PSAPs can be immediately transferred in major failures of NGCS or of the ESInet. Emergency services are critical in worst-case scenarios facing local communities, regions, and even nations. Driving that concept back to trunking, a CO with two trunks today to a data center for 9-1-1 requires two additional trunks to the mated NGCS, ideally over completely diverse facilities. Recall that it is the decision of the OSP to provide diversity, not the 9-1-1 SSP. The 9-1-1 SSP can only recommend the best solution. Given how large NG9-1-1 networks can become, this is a sensitive issue. Commission rules may change but they have not addressed this type of scenario. There were failures when initial NG9-1-1 networks were deployed. The FCC cited OSPs that failed to report the outage situations in a timely manner and created a notice of proposed rule making (NPRM) to address the situation. It covers significant recommendations but falls short of mandating OSP diversity or getting specific about 9-1-1 mandates except when reports need to be made. The penalty included fines for millions of dollars levied on the OSPs, transport providers, and 9-1-1 SSPs that failed to make a timely report with the added requirement to stipulate their plans to correct the situation going forward. Major OSPs are arguing with 9-1-1 SSPs over splitting two trunk groups in half and sending one trunk over a largely empty facility route to one NGCS and the other one existing trunk as a separate group over another to the mated NGCS. In order to balance the call load some 9-1-1 SSPs have asked for first-route and alternate-route diversity from OSPs. For example, all OSPs whose names start with a number or Alpha A through L route to NGCS A, and OSPs with names M through Z route to NGCS B, with all OSPs alternating in the opposite direction after they agree to route diversity. The goal is load-balance across the NGCS so that one set of NGCS does not remain idle until a failover. OSP trunk groups do not provide dynamic NGCS load-balanced routing. OSP switching, routing, and translations personnel have expressed a willingness to adhere to the first and alternate routing scheme; testing is needed to be sure that it works as planned before cutover. This is an area to address in connection agreements. If NGCS B is not carrying any load or a significant load daily, then the likelihood that everything works perfectly during an automatic failover or during a major network or other overload or crises is low. This is a lesson learned over 50 years of experience in designing, engineering, and managing networks.

5.5 SLAs

SLAs are needed between each OSP and the 9-1-1 SSP. Chapter 11 provides input for model SLAs.

5.6 Monitoring

Chapter 9 details OA&M requirements. Maintenance and engineering systems and processes should be in place within the 9-1-1 SSP's and each OSP's operation and engineering organizations. Monitoring needs to include as many aspects of the network and emergency services end-to-end as possible. Alerts and exception handling for each emergency call that fails are critical. This is very unique to emergency services. 9-1-1 failures are not acceptable to the public, even though commission engineering rules allow for one call in 100 to fail.

5.7 Security

Security is vital to emergency services.[15] This includes logical and physical security at each site. Emergency services are needed most when the network is under attack, locally, regionally, statewide, between states, and nationally. Cybersecurity is a major topic for all service providers especially those that are supporting emergency services. To that end, the standards for next-generation networks include the BC, which serves as a firewall for internal and external intrusions. It is not enough to install fully duplicated BCF FEs. Settings must match the size of the network, and every external and internal caller must be routed at one or more points through the firewall. The authors have seen internal attacks from disgruntled workers, or night-shift PSAP employees who plugged in devices to stream programs over the network using all the bandwidth available to process any incoming emergency calls. Before any next-generation network is turned up for service, it is important to have preprogrammed testing efforts to attack the network from every vantage point and see if it can withstand the attacks. This must be done before live traffic is placed on the network, because after that point, the exercise has the potential for real impact on service. Universities and vendors have software to provide load-generation programs and help find the weaknesses and send relevant reports to the persons in charge. While standards apply in general, no two networks are identical. Security solutions to meet those standards must be customized to the environment and the parties involved.

The federal government is involved in 9-1-1 cybersecurity efforts through the NIST, which was founded in 1901 and is now part of the U.S. Department of

15. Refer to NENA Security for Next-Generation 9-1-1 Standard (NG-SEC) NENA 75-001, Version 1, February 6, 2010.

Commerce. "NIST is one of the nation's oldest physical science laboratories. Congress established the agency to remove a major challenge to U.S. industrial competitiveness at the time - a second-rate measurement infrastructure that lagged behind the capabilities of the United Kingdom, Germany, and other economic rivals.

From the smart electric power grid and electronic health records to atomic clocks, advanced nanomaterials, and computer chips, innumerable products and services rely in some way on technology, measurement, and standards provided by the National Institute of Standards and Technology.

Today, NIST measurements support the smallest of technologies to the largest and most complex of human-made creations - from nanoscale devices so tiny that tens of thousands can fit on the end of a single human hair up to earthquake-resistant skyscrapers and global communication networks. One goal of NIST is to provide the public safety community with a better understanding of what to expect from new and emerging networking technologies, and accelerate the standardization and utilization of such technologies."[16] NIST aims to "provide the public safety community with the performance analysis tools needed to better understand emerging network technologies and facilitate: the evaluation of worst/best case network deployment scenarios, the investigation of how well new technologies support public safety requirements, the development of quantitative requirements for public safety communications, the development of next generation network standards in support of public safety communication needs. This work is part of the Public Safety Communications Research program."

5.8 NOC

Each OSP typically has a NOC. NOCs have contacts and main numbers that must be exchanged between each OSP and the 9-1-1 SSP. The 9-1-1 SSP NOC information should be provided upfront with the initial communications package. The location of the NOC and the managers' contact information are minimal pieces of information for each party to document in connection agreements. NOCs have surveillance systems critical to monitoring and alerting personnel with a need to know about events that impact 9-1-1 service. SLAs are needed between each OSP and the 9-1-1 SSPs outlining NOC responsibilities

The tier I technician handles troubles as they are known to the NOC. Typically in a major outage, a tier II maintenance person is called upon to lead the bridge and drive the outage to conclusion, create reports to management and the various commissions as required, and after closing a trouble ticket, make final reports and set up time for all parties to perform a root cause analysis (RCA) and ensure that there are improvements in processes so the event is unlikely to recur at the same site or any site with similar issues. Tier III support is provided by the vendors. If during the course of problem resolution, it is discovered that one or more vendor(s) may have a technical issue or may be upgrading equipment or software, those vendors may be need to be engaged. Vendor contracts should include an SLA with reciprocal requirements. If there is a planned activity that may affect service, the

16. https://www.nist.gov/programs-projects/public-safety-communications.

NOC should have a signed method of procedure (MOP) document ahead of the event so the impacted parties can be called if trouble is seen or occurs unexpectedly during the operation. Planned maintenance or upgrade work should only be done during authorized hours suitable to the 9-1-1 SSP NOC and all other parties with service responsibilities. In connection agreements and vendor contracts the allowable working hours for major planned activities are referred to as maintenance windows. Chapter 9 includes additional information related to NOCs, network monitoring, and a sample MOP outline.

5.9 Reporting

Commissions have variable levels of reporting requirements. When an outage or other interruption impacts emergency services, nearly all incidents are reportable to state commissions. Outages lasting over 30 minutes and impacting service potential for 30,000 customers or more must be reported to the FCC. The 9-1-1 SSP and OSPs servicing 9-1-1 access customers have reporting obligations independent of each other. It is critical to identify the number of people who reside in an area impacted by the outage, so that if and when a problem arises, the proper notifications are made in a timely fashion. Given the choice of OSP access diversity, and other challenges in identifying the number of customers impacted by an outage, the government Census has been the method for 9-1-1 SSPs to estimate customer impact for reporting purpose. Penalties and other issues, including potential lawsuits, may arise from customers if timely reports are not filed and or loss of life or property occurs during a service-impacting outage. Chapter 11 describes some legal, regulatory, and reporting requirements.

5.10 QoS

The NENA standards specify QoS parameters for the design of the ESInet and the emergency traffic that traverses the network. This includes phase, jitter, latency, and delay. Chapter 9 covers the recommended SLA QoS standards that need to be agreed upon or adjusted and documented in the OSP and 9-1-1 SSP to 9-1-1 SSP connection agreements.

5.11 Split Exchanges

Split exchanges exist where the landline telephone company switching boundaries do not match the civic public safety boundaries of the communities they serve. ILEC and RLEC OSPs have special considerations for routing emergency services calls from COs with split exchanges to the caller's designated PSAP for dispatch to first responders. An exchange is an NXX served from a wireline OSP's class 5 end office where wireline customers originate calls from a fixed address location. A split exchange is an NXX whose originating callers need to be dispatched by different PSAPS, and the customer lines are not a designated class of service. This is especially complex when PSAPs are managed by different 9-1-1 SSPs. The splitting function is performed by only one 9-1-1 SSP, called the primary 9-1-1 SSP. The primary 9-1-1 SSP is defined as the SSP serving the most lines in a split exchange. The primary 9-1-1 SSP is responsible for the analysis of and routing of most calls to their own PSAPs and routing the remaining calls over trunk groups to adjacent 9-1-1 SSPs' PSAPs for completion. CLECs are not privileged (or required) to use the same arrangements and agreements as ILECs and RLECs for split exchanges.

Figure 5.6 shows an example of a split exchange, illustrating the split exchange in the Sandwich, Illinois, frontier CO, (815) 796-XXXX, which has customers residing (i.e., split) between, three counties, DeKalb, Kendall, and LaSalle. In this example, there are three different 9-1-1 SSPs, two of which have legacy SRs and one 9-1-1 SSP with two data centers housing duplicated sets of NGCS, which include duplicated SR functions. There are two legacy PSAPs, Kendall and LaSalle, and four NG9-1-1 i3 PSAPs, two of which serve as primary and alternate PSAPs for DeKalb County and two more default PSAPs, primary Lee County and alternate Ogle County, which handle the calls in the event that the dispatch-able address is not forthcoming from the databases.

Wireline OSP-Split Exchange Basic Concepts –using Sandwich Illinois -Frontier Split Exchange

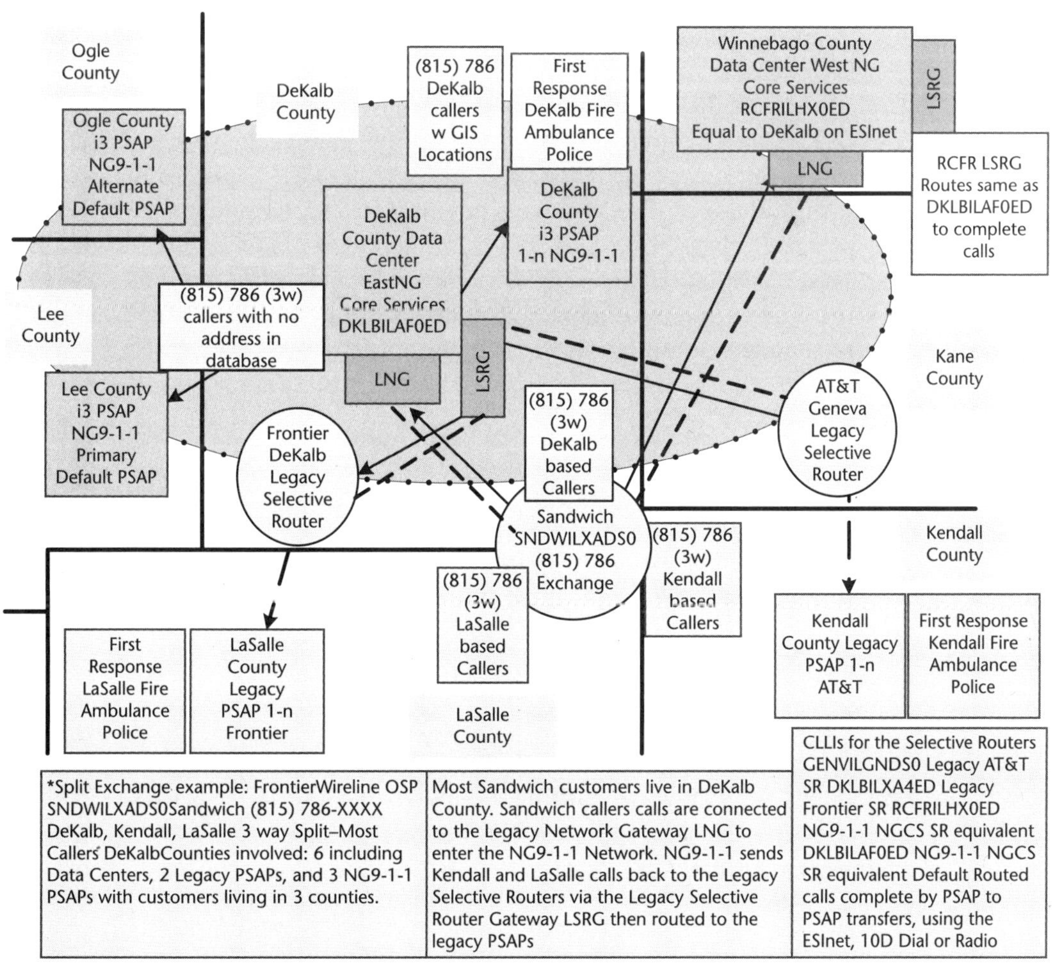

Figure 5.6 Wireline split exchange basics.

The OSP for the Sandwich 9-1-1 callers must submit to all 9-1-1 SSPs a copy of each Sandwich SO record matching the customer name and street address to each assigned 10-digit telephone number. If that information is not provided, the primary 9-1-1 SSP is not able to supply the name and customer information to the PSAP for call completion.

Approximately 10–20% of all emergency services calls originate from ILECs and RLECs. Some of these originate from split exchanges. The remaining 80–90% of all emergency services calls originate from wireless and VoIP and some CLEC wireline OSPs. NG9-1-1 standards, as described in NENA-STA-010.2-2016 [3], allow the ILEC and RLEC split exchange routing process to continue. Split exchange routing is a privilege permitted to wireline legacy OSPs, and is a complex means to accomplish the goal of getting a rapid response for all callers in an emergency. Section 5.11.1 describes the primary method of handling ILEC and RLEC split exchange routing; there are additional options described in Section 5.11.2.

5.11.1 ILEC and RLEC

The type of vendor equipment for each end office is irrelevant whether it be a circuit switch or a soft switch. An OSP may have more than one rate center (RC) and more than one NXX per RC. A RC is a regulatory term described by a state public utility commission for the origination of calls within a defined geographic footprint. Wireline callers are assigned seven-digit telephone numbers in the form of NXX-XXXX. ILECs and RLECs send traffic from split exchanges to the primary 9-1-1 SSP SR or NGCS per legal commercial agreements.[17] Split exchanges have been in use since the inception of emergency services. A component of the NGCS, the LSRG is the interface FE that enables this function to continue if legacy SRs and NG9-1-1 NGCS coexist. Figures 5.6–5.8, show how the LSRG interfaces are used to complete calls from split exchanges without PSAP to PSAP handoffs. Eventually, the need for unique split exchange routing disappears when all emergency services callers are routed to the proper PSAP using the GIS location of the caller mapped against the PSAP GIS boundaries.

5.11.2 Split Exchange Emergency Call Handling Options

There are several options available for resolving the routing of emergency calls from Split Exchanges, listed as follows.

1. Legacy 9-1-1 SSP splits;
2. Next-generation 9-1-1 SSP splits;
3. PSAP IGAs, opted in;
4. PSAP IGAs, opted out
5. One 9-1-1 SSP always splitting, no matter the line count per PSAP;
6. 9-1-1 SSP resolving splits via GIS boundary mapping when PSAPs are on the same ESInet;

17. Refer to Chapter 11 for additional information on OSP to 9-1-1 SSP commercial agreements.

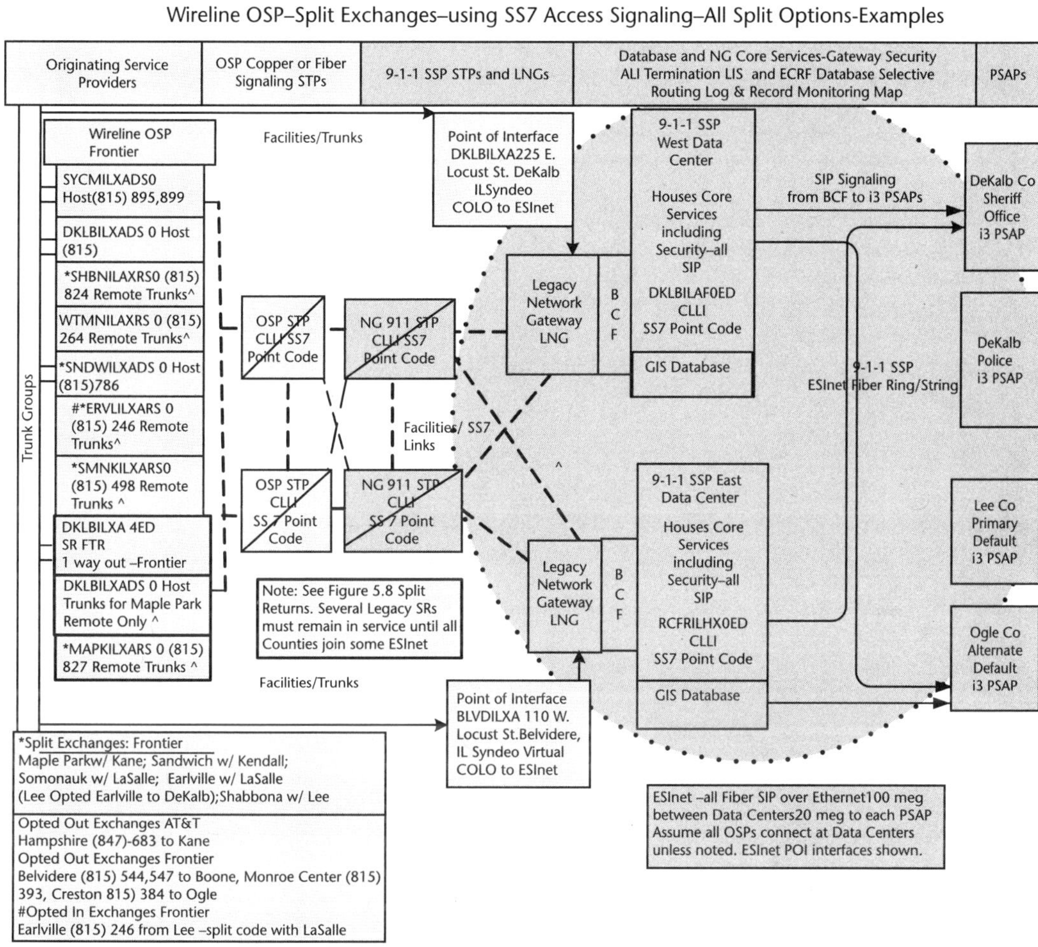

Figure 5.7 Wireline LEC with SS7 split-options basics.

7. Establishing an ESInet-to-ESInet connection, resolving inter 9-1-1 SSP splits.

5.12 Summary

OSPs are mandated by regulatory rules and statutes to provide emergency services to their customers. Few OSPs are familiar with NG9-1-1, nor do they always know where they are to terminate or understand the options to terminate their calls. Regulatory agencies govern emergency services, and often they do not know how NG9-1-1 works. The OSPs are responsible to both the FCC and to individual state public utility commissions for service quality. 9-1-1 SSPs must work cooperatively with each OSP to reach agreement on the volume of calls that is generated and how the network is sized and managed to meet the commissions' regulatory requirements. There are many standards for the design and implementation of emergency services networks. The standards for interconnection between OSP networks and the emergency next-generation networks have many options. This book strives to

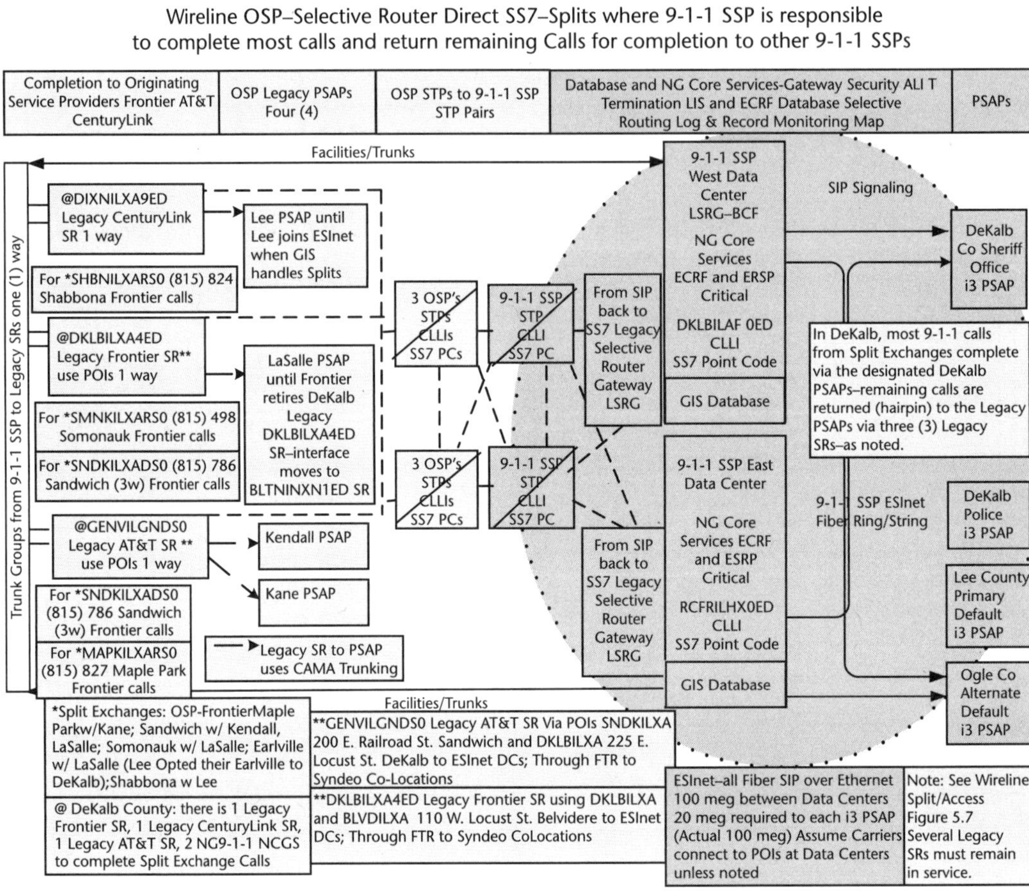

Figure 5.8 Wireline LEC call completion through next-generation and legacy SRs.

bridge communications and terminology gaps and address the options from a practical perspective. Cooperation between OSPs and the 9-1-1 SSPs is mandatory to maintain a high level of service in what appears to be a very long transition period, since migration is not mandatory and, for the most part, has no time frames. Regulatory approval is required to become a certified carrier or OSP in most states.

Selected Bibliography

NENA-STA-010.2-2016; Full Name: Detailed Functional and Interface Standards for the NENA i3 Solution Document, Originally 08-003. www.NENA.org.

CHAPTER 6

Enterprise Customers

6.1 Overview

This chapter addresses large enterprise customers that require special consideration when planning for NG9-1-1. Enterprise and multiline telephone systems[1] (MLTSs) are terms used when referring to such customers. Residences and small businesses often have multiple lines of telephone service and various types of service including wireline, wireless, and VoIP. This chapter addresses enterprises large enough to require unique requirements for 9-1-1 service.

This chapter introduces the term private switch (PS) 9-1-1 (PS/911), which is the type of equipment in place in typical enterprise environments. PS/911 includes private branch exchanges (PBXs) switches, or the centrex business telephone service offered by landline local exchange OSPs. Enterprises included in this chapter are typically large businesses, universities, and government. An enterprise's size and locations determine its connectivity to the emergency services platforms. An enterprise network[2] is defined as a large network connecting major points in a company or other organizations and not part of the network's public telecommunications infrastructure. In the United States, there are varying rules and regulations on a state-by-state basis governing how an enterprise should complete calls to emergency services. Federal rules and regulations explicitly govern enterprise VoIP OSPs and enterprise PS/911 requirements for reaching emergency services including expectations for caller location accuracy. This chapter includes references to NENA requirements, FCC regulations, and sample state commission requirements. Note that hotels are an important type of enterprise customer. A tragedy that occurred in a hotel was the motivation for Kari's Law[3], a federal law that became official on September 26, 2018. The law was named for a Texas woman, Kari Hunt, who was murdered in a Texas hotel by her ex-husband while her daughter dialed 9-1-1 repeatedly and unsuccessfully, not realizing she needed to dial a prefix before the

1. MLTS customers are generally serviced by PBX and centrex systems or switch types. (NENA Master Glossary of 9-1-1 Terminology NENA-ADM-000.22-2018, 04/13/2018.)
2. An enterprise network is a large network connecting major points in a company or other organizations not part of the network's public telecommunications infrastructure. (NENA Master Glossary of 9-1-1 Terminology NENA-ADM-000.22-2018, 04/13/2018.)
3. https://www.fcc.gov/news-events/podcast/personal-story-behind-karis-law.

digits 9-1-1. Kari's Law was passed by several states before becoming federal law. Implementation is in progress but not complete across the United States. Interviews with a focus on enterprise NG9-1-1 were used as a basis for Chapter 6.[4]

6.2 Enterprise Regulatory Requirements

NENA standards provide insight as to the reasonable limits for enterprises to send 9-1-1 calls via direct connections to emergency services legacy SRs, or NG9-1-1 NGCS, to reach the designated PSAPs. Enterprises below the limits with fewer lines or locations may, according to NENA, send their 9-1-1 calls through the OSP network with all other calls as described in Chapter 5. Per NENA guidelines, "Current acceptable limits suggest that areas of no more than 40,000 square feet with no more than 49 telephone stations on a single floor in a single building may be assigned an Emergency Response Location (ERL). One telephone number within this area is established as the Emergency Location Identification Number (ELIN). The MLTS Telephone System must be programmed to cause each telephone station located within this given area to send the ELIN as the ANI when a 9-1-1 call is made."[5] Sections 6.1–6.4 address requirements and options for enterprise 9-1-1 call flow. Location services are elaborated as follows: With respect to an ERL, NENA standards state: "It is, therefore, required that each Enterprise provide accurate location information or Emergency Response Location (ERL), for every telephone capable of dialing 9-1-1 and identifying its number or ELIN to the 9-1-1 network." The standard further states: "Every telephone capable of dialing 9-1-1 and identifying its telephone number, its ANI or Emergency Location Identification Number (ELIN) must have a location or ALI record in the 9-1-1 database to identify the caller's location."

Section 6.11 addresses location services for enterprises with respect to 9-1-1.

Within the United States, states with explicit requirements for enterprise network connectivity to 9-1-1 include: Alaska, Arkansas, Colorado, Connecticut, Florida, Illinois, Kentucky, Louisiana, Maine, Maryland, Massachusetts, Michigan, Minnesota, Mississippi, New Hampshire, New Jersey, Oklahoma, Pennsylvania, Tennessee, Texas, Utah, Vermont, Virginia, and Washington.

The following examples show that states have their own interpretation of the standards. Enterprises must follow state, federal, and international regulations.

Illinois:

> Private residential switch service providers must identify the telephone number, extension number, and the physical location of a 911 caller to the PSAP. Private business switch service providers must provide ANI and ALI data for each 911 call, and must not require the dialing of an additional digit prefix[6] (systems installed after

4. William, M., NENA ICE Chairperson, Personal Communications, May 7, 2019. Bill Mertka has extensive NG9-1-1 experience with a major focus on the enterprise.
5. NENA-06-003 Revised August 2004 Private Switch (PS) E-9-1-1 Database Standard, 2.1.3.3.
6. Additional references for Kari's Law: https://www.congress.gov/bill/114th-congress/house-bill/4167; https://www.fcc.gov/document/pshsb-announces-comment-date-nprm-implementing-karis-law.

July 1, 2015); the level of detail required for ALI data, exemptions and guidelines to establish a private emergency answering point are outlined.

Learn more at the following sites:

- http://www.ilga.gov/legislation/ilcs/ilcs3.asp?ActID=741&ChapterID=11;
- http://ilga.gov/commission/jcar/admincode/083/08300726sections.html;
- http://ilga.gov/commission/jcar/admincode/083/08300727sections.html.

Maine:

Residential MLTS providers must deliver a distinct ANI and ALI for each living unit to the PSAP. Business MLTS providers must deliver ANI and ALI to the PSAP; specific ALI data requirements are outlined." Maine also includes requirements for hotels/motels, as well as exemptions and guidelines to establish a private emergency answering point. Any public or private entity that installs or operates a multi-line telephone system ensures that it is connected to the PSTN in such a way that 911 can be dialed without requiring any prefixes.

Learn more:

- http://www.maine911.com/laws_rules/rules.htm;
- http://www.mainelegislature.org/legis/bills/getPDF.asp?paper=SP0271&item=1&snum=128.

Rules may be found on individual commission websites and are described on several third-party providers' websites. Third-party providers compete to assist enterprises to meet their obligations to provide emergency services. NENA documents mention providers such as West Public Safety, Comtech TCS, RedSky, and Level 3. It is incumbent on each enterprise and the responsible 9-1-1 SSP for an area to know about and abide by the FCC and state obligations to assist their callers with reaching emergency services in a timely manner.

The FCC promulgated rules for VoIP providers as it pertains to emergency services and general rules for all types of OSPs for location services. For complete information go to the FCC[7] website.

FCC VoIP 9-1-1:

To reduce possible risks to public safety, the FCC requires interconnected VoIP providers to: Automatically provide 911 service to all customers as a standard, mandatory feature. VoIP providers may not allow customers to option out of 911 service. Obtain a customer's physical location prior to service activation and provide one or more easy ways for customers to update the location they have registered with the provider if it changes. Transmit all 911 calls, as well as a callback number (CBN) and the caller's registered physical location, to the appropriate emergency services call center or local emergency authority. Take appropriate action to ensure customers have a clear understanding of the limitations, if any, of their 911 service.

7. https://transition.fcc.gov/cgb/consumerfacts/voip911.pdf.

They must distribute labels warning customers if 911 service may be limited or not available and instruct them to place the labels on or near equipment used with VoIP service. Obtain affirmative acknowledgement from all customers that they are aware of and understand the limitations of their 911 service.

Ensure that a 911 call is routed to the appropriate PSAP in areas where emergency service providers are not capable of receiving or processing the location information or call back numbers not automatically transmitted with 911 calls. VoIP service providers that do not fully interconnect with the PSTN are not currently required to comply with the FCC's 911 and E911 rules.

The NENA standards contain references detailing the requirements for large MLTS entities. Be aware that there is not a lot of published experiential data relative to the large enterprise and the NG9-1-1 transition, because by and large those enterprises are still routing their emergency calls first to the legacy SR, even where a state region or PSAP has been officially converted to NG9-1-1.

Whether the enterprise is small or large, 9-1-1 calls from it originate from a PBX or a centrex that is a part of an OSP's switching system. In the case of the small enterprise, the number of lines in the MLTS entity is small enough to allow the originating OSP to carry its traffic as discussed in Chapter 5 using location services rules that enable dispatch to the site without undue delay or confusion. The 9-1-1 SSP's databases allow for additional fields to describe building, floor, and room numbers with full addresses for the first responders to locate the emergency. The small MLTS entity's calls flow exactly as any residential call from that OSP.

Large enterprises, on the other hand, have many more lines and a larger footprint, possibly one or more buildings using a single MLTS phone system across a large geographic area spanning multiple PSAPs. The types of PBX and centrex equipment may vary. Many entities have migrated to business VoIP services. Some OSPs have migrated their large enterprise customers from circuit-switched, centrex TDM offerings to SoftSwitch centrex offerings. There are several NENA standards,[8] many predating NG9-1-1, that support the requirements at a PBX or a centrex to route calls successfully to emergency services. One area they all have in common is the option for dedicated trunking to either an E9-1-1 SR or the NG9-1-1 NGCS. Another common area is a requirement for a DBMS to improve location services. Enterprises are responsible for sending their location information to a 9-1-1 DBMS. In many cases, enterprises outsource those responsibilities to either OSPs or third-party providers that specialize in the ever-expanding options for both accessing 9-1-1 networks and for providing accurate database information.

6.2.1 Rationale for Enterprise Cutover Phases

In any area there may be dozens if not hundreds of OSPs to connect directly into a NG9-1-1 network. Planning, engineering, testing, and cutting OSPs to their new

8. Examples of pre-NG9-1-1 MLTS standards: NENA 06-502, Version 1, October 25, 2008, Industry Common Mechanisms for MLTS E9-1-1 Caller Location Discovery and Reporting Technical Information Documents (TID); and NENA 08-752 Issue 1, December 21, 2006, NENA Technical Requirements for Location.

configuration are significant undertakings and often are broken down in phases. Each NGCS cutover is going to be determined by what makes sense for the area. The geography and the legal and regulatory constraints of the area are primary considerations. In terms of enterprise cutovers, the number of enterprises needing connectivity into the NGCS can be significant. It is suggested with initial cutovers that the large enterprises remain behind the legacy SRs until the core NGCS is tested and ready for any 9-1-1 calls. Phase 1 of access cutovers should include the OSPs with connectivity to legacy SRs. Phase 2 of access cutovers should include a change of access for large enterprises that must route directly to the NGCS bypassing the OSP switches for 9-1-1 services. There may be exceptions, including enterprises with a presence on the ESInet and with switch routing functions capable of routing 9-1-1 calls directly to the NGCS in phase 1.

NENA recommends that each line of an enterprise be tested for 9-1-1 call arrival at the correct PSAP with location accuracy. OSP cutovers require a sample of lines and classes of service to be tested before cutover. The difference is significant to the success of the callers, the 9-1-1 SSPs, the PSAPs, and all staff personnel doing the conversion process. See Chapter 7 for more information.

Enterprises require a different legal arrangement or contract for 9-1-1 connectivity than OSPs. The existing 9-1-1 SSPs have been reluctant to divulge the contacts for those contracts for the new 9-1-1 SSP to begin negotiation. There is no state, federal, or NENA database or registry to identify the enterprises with such a contact. Refer to Chapter 11 for further discussion.

6.2.2 Phase 1 Network Connectivity Options

Known deployments have the enterprise customers first routing to the E-9-1-1 legacy SR, which is nearly always a TDM circuit switch. The legacy SR sends the calls to the NG9-1-1 ESInet where they are handled by the NGCS. To reach the PSAPs, connectivity can be through the LNG or the SBC. The authors are not aware of any enterprise that has been directly connected to a NG9-1-1 NGCS using SIP.

Figure 6.1 reflects what this book calls phase-1[9] methods implemented by early adopters to enable enterprise connectivity to NG9-1-1 networks. Phase 1 is defined for transition of the enterprise customers by directly routing 9-1-1 to NG9-1-1 networks for selected enterprise customers at the same time as the OSPs are cutover. In subsequent phases, the remaining majority of enterprise customers may be cut over into the new NG9-1-1 configuration one-by-one.

There are two sample phase 1 direct-to-9-1-1 enterprises depicted in Figure 6.1: (1) a government enterprise TDM PBX using ISDN-PRI connected directly to the NG9-1-1 network through the LNG without diverse access, and (2) a business enterprise VoIP IP PBX using SIP connected directly to the NGCS through the BCF with diverse access. The university IP PBX with ISDN-PRI remains connected

9. Each 9-1-1 SSP needs to assess the criticality of moving the enterprise customers to a new configuration. In a project involving multiple phases where different carriers are cut live over a period of time because of the number of OSPs. The first phase (phase 1) is considered by the authors as the phase to transition any enterprise customers that are large enough or significant enough to directly route their 9-1-1 traffic directly to the NG9-1-1 networks similar to a OSP. The majority of enterprise customers are targeted to cut over at the same time that the OSP(s) that serve them are cutover.

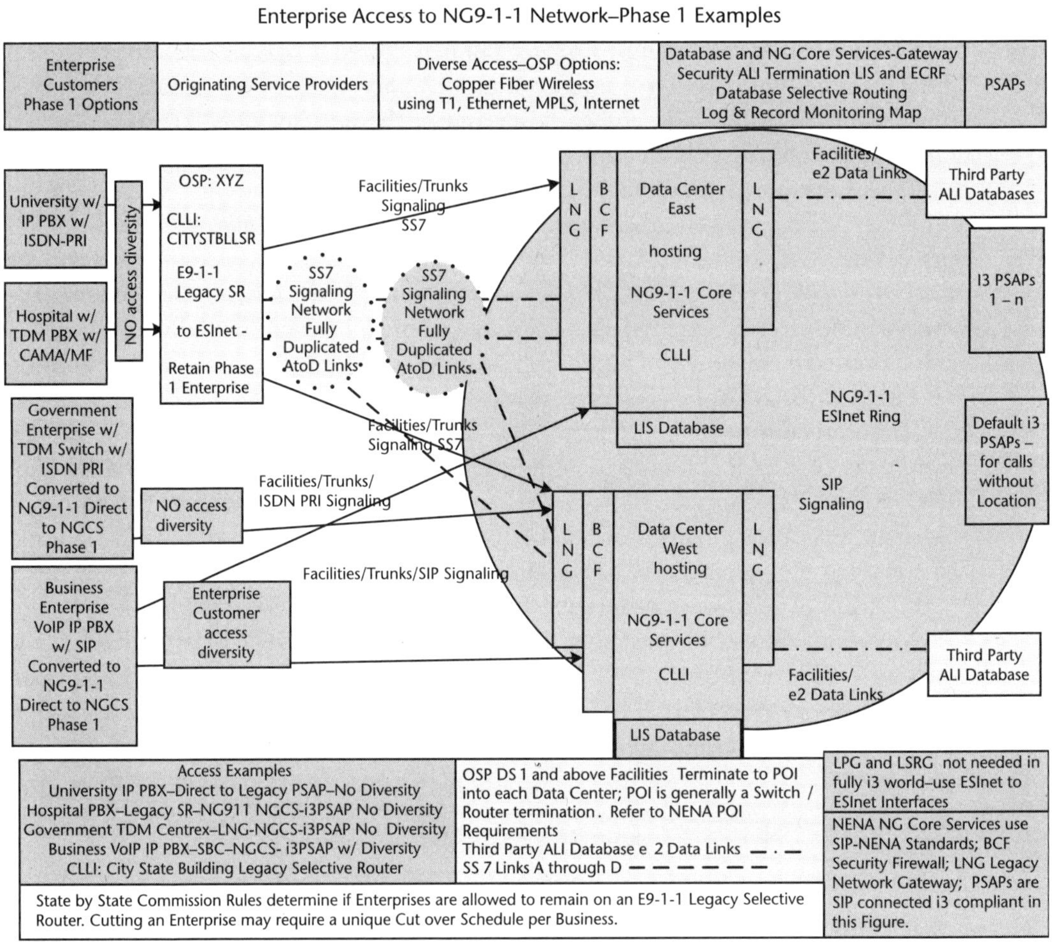

Figure 6.1 Examples of enterprise access to NG9-1-1 phase 1.

directly (no diversity) to an E9-1-1 legacy SR and that E9-1-1 legacy SR is using diverse SS7 signaling and trunks to the NGCS through the LNGs. The hospital TDM PBX with CAMA/MF remains connected via the E9-1-1 legacy SR and then to the NGCS via the LNG in the same way as the government PBX.

Figure 6.1 implies that some enterprise 9-1-1 contracts are in effect; not all PS/911 enterprise customers are ready to directly connect to NG9-1-1 networks. With respect to Figure 6.1, it is important to note that enterprise PS/911 customers, OSPs, and third-party aggregators choose their own switch types, transport carriers, signaling, and access diversity. While access diversity is highly recommended, it is the business decision of each enterprise, in concert with its OSP and/or third-party provider, to decide how to connect to the NG9-1-1 network. The type of signaling chosen is important to identify the proper POI for connection. Continuous evolution in technology will occur, thus requiring ongoing changes to the end-to-end connectivity. Routing in the new NG9-1-1 DBMS inside the network or using an overlay third-party solution directs originating callers to the correct PSAP or PSAPs. The permutations and combinations of legacy SRs and database management solutions and PSAPs are large. The actual information to set up the routing and testing is a complex effort of equal importance to arranging the network portion.

In many jurisdictions, 9-1-1 SSPs must provide annual 9-1-1 network diagrams to the state commissions that oversee their 9-1-1 customers and PSAPs. Seldom do those diagrams reflect enterprise trunking or routing, nor do they include the names of the enterprises that are routed to an E9-1-1 legacy SR. There are private contracts that are challenging to obtain if the new NG9-1-1 SSP is not the same as the former 9-1-1 SSP. This is a legal and regulatory issue for design engineering and implementation. Information related to the volume of anticipated traffic from each enterprise and the capacity of the facilities and trunking that are required to carry it must be included from the beginning of the NG9-1-1 migration plan. A plan for how enterprise location and other database information is implemented is necessary from the start of a project. Delays in implementing NG9-1-1 networks may occur if this information is not addressed upfront in the RFP and planning intervals.

6.2.3 Enterprise Access to NG9-1-1 Network

In Figure 6.2, the legacy SR has been eliminated from the call flow illustrated in Figure 6.1. This eliminates a SPOF from the end-to-end emergency services architecture. As of this writing many of the legacy SRs are manufacturer-discontinued. Until enterprises are routed to either the NGCS through an LNG or BCF to reach an i3 PSAP, the legacy SRs cannot be decommissioned.

In Figure 6.2's examples, the enterprises have not changed their choice of equipment: IP or TDM PBX or centrex. Note that changes to the equipment at the enterprise are ongoing and must be monitored from start to finish of any project and beyond.

Consider the following examples: The university enterprise 9-1-1 is directly connected to the NG9-1-1 network via the LNG to the NGCS, which may or may not be in the same location as the old E9-1-1 SR. The hospital enterprise has migrated to connect directly to the LNG using the same signaling it had before, thus requiring facilities to be terminated into the LNG. The hospital chose no access diversity. If this hospital was the last enterprise or OSP requiring the legacy SR in the architecture, that legacy SR could be retired from service. As the network migrates away from legacy signaling, SS7 network access and additional costs can be reduced. To be clear, LNGs are transitional functional elements. The timeframe for elimination is long-term since there are no mandates for end-to-end SIP or ISDN-PRI signaling for accessing NG9-1-1 networks at large. Section 6.11 covers enterprise location and database services.

6.3 Connectivity Options

This section addresses methods and options for connecting a large enterprise to the NG9-1-1 network.

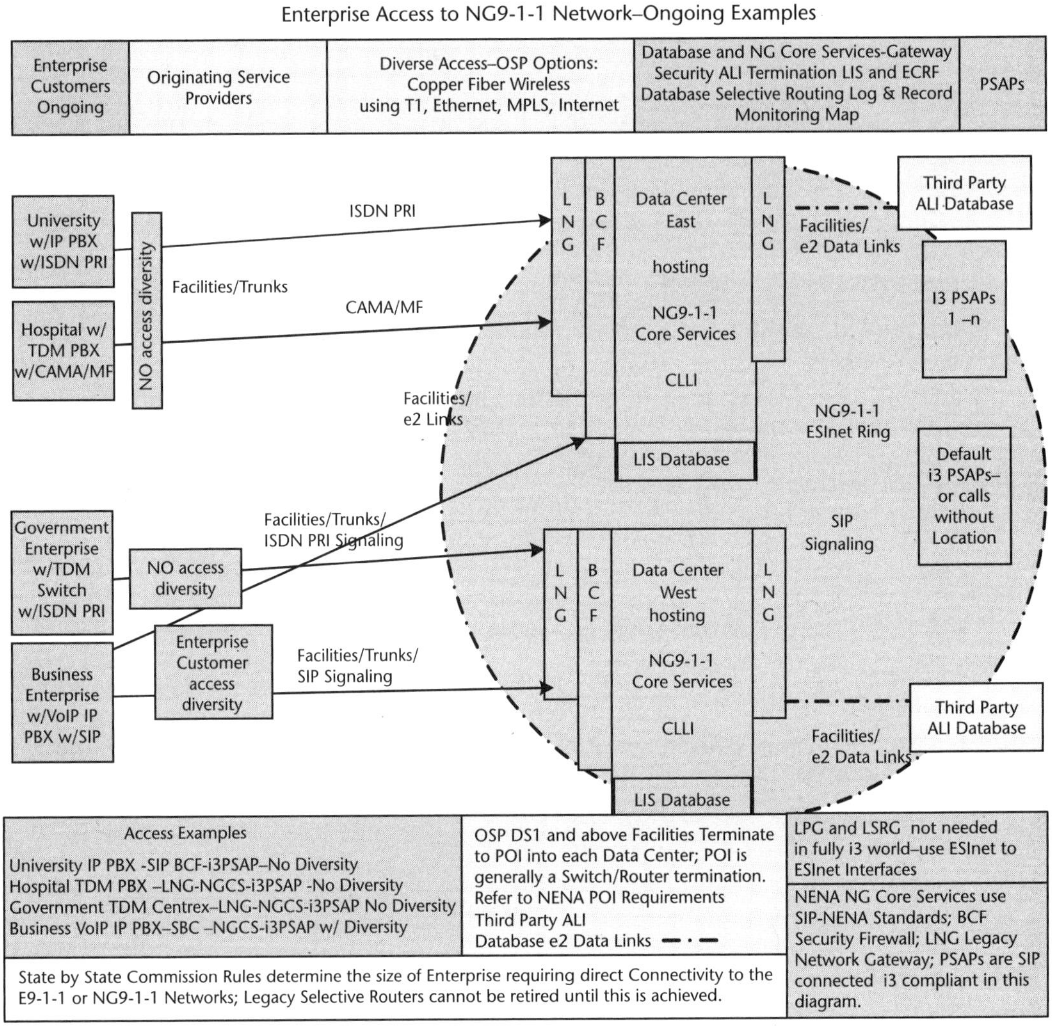

Figure 6.2 Enterprise access to NG9-1-1 network—ongoing.

6.3.1 Facilities

To set the stage for a discussion about facilities between the enterprise MLTS providers or operators and the NG9-1-1 network, it is essential to identify the points of interface or POIs and the responsibilities of each party. NENA has a formal reference that covers the OSP interfaces and enterprise interfaces to connect to NG9-1-1 networks most efficiently to reach the appropriate PSAPs.[10] Per NENA standards: "For OSPs and Enterprises there are both physical and logical interface demarcation points. Physical demarcation points are between the actual hardware elements within the NG9-1-1 network, and between the NG9-1-1 network and the originating OSP or Enterprise networks, and the PSAP facilities. Logical demarcation points are determined by interfaces between FEs within an NG9-1-1 network and

10. NENA Potential Points of Demarcation in NG9-1-1 Networks Information Document NENA-INF-003.1-2013, DSC Approval: 01/22/2013, PRC Approval: 03/01/2013, NENA Executive Board Approval: 03/21/2013.

those devices and/or applications that interact with the NG9-1-1 network." With respect to the data models, the physical demarcation points indicate where cables are connected. The logical demarcation points show the flow of data communications protocols between cooperating software functions. "A demarcation point is a mutually-defined boundary dividing one area of responsibility from another. A demarcation point has no meaning unless all stakeholders agree on its location, such as if that location is specified by industry standards or regulations, and so it is critical that a demarcation point is established through consensus."[11]

"The parties responsible for network elements or functions on one side of a demarcation point are responsible to ensure that those elements or functions are operating properly and to specification up to, and not beyond, that point. This responsibility does not necessarily indicate ownership, as elements of the network may be leased or provided by a Third-Party. The party responsible for each element may be expected to fund implementation, maintenance, upgrades, and any other costs necessary to ensure proper operation. The party responsible may elect to own and operate a network element it is responsible for or may elect to contract with a Third-Party for those services. There are operational impacts associated with any newly identified type of demarcation. These newly identified points have new and different interfaces for physical interconnection, data."

6.3.2 Interconnection

Enterprises can connect in many different ways. Chapter 5 discusses connections through an OSP. Small enterprises connect to a nearby OSP switch and when that OSP migrates to a NG9-1-1 network the change should be nearly invisible to the small enterprise. Line testing is recommended and should be managed between the OSP and the NG9-1-1 SSP as part of the planning, implementation, and testing process. NENA standards recommend testing that includes all classes of service including PBX and centrex services to be sure there are no end-to-end compatibility problems and that the customers, whether they be single-line businesses or larger businesses, are still routing to the correct PSAPs before, during, and after cutover. If there are changes in location databases, that is a negotiated responsibility for the OSP in concert with the enterprise and, where required, the enterprise's third-party provider.

Where large enterprises are directly connected to existing E9-1-1 legacy SRs, there are added decisions to be made. Large enterprises may choose to connect in multiple ways for NG9-1-1 service. The following are options: (1) remain behind a legacy SR (for phase 1) where allowed by the commission and/or or 9-1-1 SSP, and make no change, or (2) move to a direct connection to an NG9-1-1 NGCS architecture to the proper PSAP(s) through an ESInet. To fully understand the options for direct connection to either the NGCS, the enterprise's project manager needs a description and location and the identification of data centers or COs connected to the ESInet that house the NGCS for the NG9-1-1 network and the addresses and names of the E9-1-1 legacy PSAPs or new NG9-1-1 i3 PSAPs. Note that many PSAP consolidations are taking place and that the targeted dates and locations need

11. NENA Potential Points of Demarcation in NG9-1-1 Networks Information Document NENA-INF-003.1-2013, March 21, 2013.

to be clear to the parties in negotiation. The addresses and CLLI for the enterprises, NGCSs, and i3 PSAPs and the options for connection are equally important. The 9-1-1 SSP needs to specify the POIs along the ESInet where enterprises can connect. The POIs must be secure, powered, grounded, and ready to connect to enterprise equipment and provide switches and/or routers to allow the calls to flow in any direction around the ESInet should any path along the ESInet fail for any reason. NG9-1-1 standards for connecting facilities and using switches and routers allow for nearly every transport technology: fiber, copper, wireless, microwave, and satellite. While Internet access is a NENA-allowable option, most 9-1-1 SSPs provide secure, dedicated access. Each enterprise decides whether to interconnect diversely regardless of the chosen means to connect for NG9-1-1 services and regardless of the decisions related to transport or other technology. The enterprise project manager needs to be assured that if one of the data centers, POIs, or interfaces fails, the ESInet and NGCS design allows their calls to automatically reroute to complete to a designated primary or alternate PSAP. Most NG9-1-1 NGCS equipment vendors and 9-1-1 SSPs or their i3 PSAPs will not accept connections below a T1 or E1 level. Many Ethernet providers do not sell access below 10 megabits of bandwidth over an IP backbone network. Note that although it is ideal that the ESInet is built on a fiber ring, that ring should not be a collapsed[12] ring. Enterprises decide how and where to terminate their facilities. The enterprise should ask to review the ESInet design to determine if the architecture is built to standards. Each enterprise determines how much access bandwidth it requires. The NG9-1-1 network accepts the emergency calls from each enterprise and routes them to the correct PSAPs. ESInets with duplicated NGCS add reliability benefits for enterprises that currently have to complete through legacy SRs today without SR redundancy. Enterprises may be able to connect at a regional, state, or multistate level. For example, many wireless OSPs are connected to a whole state using only two POIs. Such connections are technically efficient and cost-effective. Once transport bandwidth is increased for access, a larger number of calls can route into the NGCS more efficiently. To properly size the ESInet and transport to the PSAPs, it is essential to know how much traffic is forecast to route from any particular enterprise in the busiest hour of the busiest season projected for 9-1-1 services. Enterprise peak calling business hours may vary significantly from one to another and may have little in common with the business hours and demands of the legacy SR where the enterprise's traffic routes to and from today. The 9-1-1 trunk group busy hour and busy season is typically not the same as the overall network busy hour and busy season. Traffic studies may be required if an enterprise has no historical records to assist the 9-1-1 SSP. The 9-1-1 SSP needs to set the incoming secure firewall or BCF at an appropriate number of queries per second to allow all the good calls to go through while preventing an overload to any NGCS component and/or terminating PSAP. The number of simultaneous calls and the average time per call dictate the minimal traffic to let through the BCF. The discussion about traffic load from an enterprise and the impact on the NGCS architecture is an exercise to determine total demand on the ESInet and the

12. A collapsed ring refers to a ring architecture where the physical fibers used for connectivity both reside in the same physical bundle or buffer tube. The result of this is that typically in the event of a fiber cut, the entire ring will be taken down.

NGCS. That is useful for the engineering of the total network infrastructure. It also helps determine the time of day and day of week to cut an enterprise.

State rules vary on whether to allow emergency calls to remain in queue to a PSAP or to let callers hear a fast busy or announcement so they know to redial. Testing the BCF is another important function before adding live traffic to any NG9-1-1 network. There is no way for the PSAP or a new 9-1-1 SSP to know if there is pent-up demand that could be released when the new network design is implemented. Networks have become overloaded when unlimited wireless demand has full access to limited positions per PSAP, for example. As the network is cut over for new services, the 9-1-1 SSP should consider adding enterprise customers incrementally and monitoring each closely for service impact.

When using T1s or E1s over copper, a full T1 or E1 should be ordered. If the transport connection is via Ethernet, no less than 10 megabits of capacity are usually available for purchase. Each enterprise needs to have a legal contract with the 9-1-1 SSP that must include the type of transport, the desired bandwidth, and the decision for the option to choose access diversity. The details of a contract ought to be more comprehensive than facilities. It is important to enter into contact negotiations early so that the decisions benefit all parties.

6.4 Signaling

Enterprise to NG9-1-1 originating signaling options for emergency services include SIP, ISDN-PRI, and CAMA/MF. SIP is preferred, followed by ISDN-PRI, and enterprise PBXs and centrex switches are often capable of using these signaling methods. If the enterprise uses an IP PBX or SoftSwitch centrex switch, internal calls use SIP signaling. This does not mean however, that calls to phones outside the enterprise are SIP calls. For the enterprise to make SIP calls directly to the ESInet and the NGCS, the enterprise needs facilities that allow it to send SIP call requests over facilities to a BCF, the gateway to the ESInet and the NGCS. This type of facility is referred to as a SIP trunk. Since the enterprise may not have experience using SIP trunking to the outside world, these connections need to be discussed early in the planning and design process and thoroughly tested with each IP-PBX and Softswitch centrex manufacturer to be sure that their version of SIP is compatible with NG9-1-1. Information about the SIP capabilities required for NG9-1-1 compatibility can be found in Chapter 3. Note that some enterprises may have already chosen to send their off-net calls to a SIP OSP over a SIP trunk group. Most of those calls are routed to the PSTN phones via the OSP's PSTN gateway. Routing 9-1-1 calls via a SIP service provider, described in Chapter 3, is different from the direct connection discussed here.

There are good reasons to opt for a SIP signaling connection to the emergency services network: (1) the backbone of the NG9-1-1 network uses SIP signaling; (2) SIP call setup is nominally faster than analog and TDM methods, so that the caller can experience a faster and more efficient end-to-end call experience; (3) the NGCS have interfaces to the NG9-1-1 databases for location and additional data queries either within the NGCS databases or to connections to third-party DBMS services; (4) the i3PSAPs are connected to the ESInet and are also protected by the BCF.

Today most enterprise 9-1-1 trunks use dedicated ISDN-PRI or CAMA/MF signaling even though their other calls are routed using whatever optimal interface and signaling agreements are offered by their OSP. While it is possible per NENA standards to continue using the CAMA/MF signaling, it is not an optimal solution. CAMA/MF trunks were needed when 9-1-1 was introduced to allow the PSAP to obtain enough caller ID information to allow proper PSAP routing by the OSP's legacy SR. There were no other viable signaling options 50 years ago; NENA standards grandfathered CAMA/MF signaling for NG9-1-1. There is a limited amount of CAMA/MF testing equipment to perform routine tests. The call setup time for a CAMA/MF connection is longer than for modern signaling; calls using SIP, SS7, or ISDN-PRI have faster call setup times than those using CAMA trunks. Emergency callers are accustomed to hearing ringing tones quickly on nonemergency calls that use SIP, SS7, or ISDN-PRI; they often hang up and try their call again if they think an emergency call is not progressing. This can cause loss of valuable time for an emergency caller to be connected and speak to the PSAP call-taker.

6.4.1 Location Signaling

Depending on the enterprise, connectivity to the i3 PSAPs may include access back through the ESInet to obtain the location information, or they may be connected directly to the location databases the enterprise maintains. The standards-based signaling between the PSAP and NGCS to external third-party databases is via e2 datalink signaling, which is described in Section 6.11. Newer solutions are emerging as discussed in Chapter 10. Enterprise contracts must stipulate the routing for emergency calls and databases including the access signaling specifications and database management signaling and location choices. All legacy signaling interface calls, ISDN-PRI, and CAMA/MF must arrive first at the LNG for signaling conversion to SIP before being passed to the NGCS for completion to the correct PSAPs. LNGs are transitional functional elements to convert legacy signaling to SIP signaling required by the NGCS and i3 PSAPs.

6.4.2 Jurisdictional Boundaries

If there is a PSAP outside the jurisdiction of the 9-1-1 SSP that must be reached, there are options for the enterprise to retain several trunk groups, one to the older E9-1-1 networks and one to the NG9-1-1 networks or negotiate an ESInet to ESInet connection to reach those PSAPs. This requires extensive testing since more than one 9-1-1 SSP may be involved, and the choice of 9-1-1 equipment is unique to each 9-1-1 SSP. All the equipment should have been tested at least once at a NENA ICE. ESInet-to-ESInet test plans and results have not been made publically available. NENA has proposed extensive official public compliance testing through a third-party laboratory; compliance testing requirements have not been released for bid as an RFP as of December 2020.

Enterprise contacts are important to begin contract negotiation before enterprise direct connection to NG9-1-1 can take place. Signaling options are an important component of the negotiation. See Chapter 11 for further discussion.

6.5 Trunking

The 9-1-1 SSP needs to know the number of trunks each large enterprise will provide into the NG9-1-1 architecture. In general, enterprises design the trunk group sizes based on the anticipated emergency services traffic. For OSPs, most state and federal commissions require that no more than one emergency call in 100 is blocked in the busiest hour of the busiest rolling 20-day busy season of the year. This is called P.01 GoS. The P comes from Poisson[13] traffic engineering theory. NENA suggests no less than two trunks be provided for access by any enterprise for local access, wireline or wireless switch into the 9-1-1 network. Traffic demand using engineering formulae determines the number of trunks required to meet the required GoS. The 9-1-1 SSP typically cannot see the access side of the network and thus accepts the number of trunks required from each OSP or enterprise to reach all PSAPs. As stated earlier, it is the decision of the enterprise to provide diversity, not the 9-1-1 SSP. The 9-1-1 SSP can and should work with enterprises to determine the best solution. Given how large NG9-1-1 networks can become, this is a sensitive issue. Commission rules may change but they have not addressed this type of scenario. State and FCC commissions fall short of mandating OSP or enterprise diversity or getting specific about 9-1-1 mandates except when OSP reports need to be made. If and when enterprises connect directly to NG9-1-1 networks or PSAPs, there are no formal requirements for the PSAPs or the 9-1-1 SSP to notify enterprise customers if there is a failure. There is one opportunity to include a requirement for mutual notification, and that is through legal agreements and contracts.

6.6 SLAs

SLAs are needed between each enterprise and 9-1-1 SSP. There is no model SLA for enterprise customers available. A brief review of a typical OSP to NG9-1-1 SSP SLA may yield talking points. See Chapter 11.

6.7 Monitoring

Chapter 9 details OA&M requirements. Maintenance and engineering systems and processes should be in place within the 9-1-1 SSP's organizations and within each enterprise's operations and engineering organizations. Each enterprise is unique.

13. Telecommunications traffic engineering, teletraffic engineering, or traffic engineering is the application of traffic engineering theory to telecommunications. Teletraffic engineers use their knowledge of statistics including queuing theory, the nature of traffic, their practical models, their measurements, and their simulations to make predictions and to plan telecommunication networks such as a telephone network or the Internet. These tools and knowledge help provide reliable service at lower cost.
The field was created by the work of A. K. Erlang for circuit-switched networks but is applicable to packet-switched networks, as they both exhibit Markovian properties and can hence be modeled by, for example, a Poisson arrival process.
The crucial observation in traffic engineering is that in large systems the law of large numbers can be used to make the aggregate properties of a system over a long period of time much more predictable than the behavior of individual parts of the system.

Some have 9-1-1 provisioning and maintenance responsibilities in a unique group, while others may have telecommunications and/or IT group(s) that work together to perform those functions. Monitoring needs to include as many aspects of the network and emergency services end-to-end as possible. Alerts and exception handling for each emergency call that fails is critical. This is a unique exception for emergency services. 9-1-1 failures are not acceptable to the public. The authors are not aware whether any commissions regulate enterprises with respect to blocking criteria or reporting of outages. With the focus on active shooter drills and incidents on school campuses and at business and government locations over the years, it is apparent that emergency services are an important component of safety and security. Dedicated security organizations may be in a position to monitor emergency services and/or be appraised of an outage impacting access to emergency services. Location services are critical. Consider monitoring, notification and alerting processes when negotiating enterprise contracts.

6.8 Security

Security is vital to emergency services. The security of both the enterprise network and that of the NG9-1-1 network must be considered and the secure flow of information between the enterprise, NG9-1-1 network, and the PSAP must be verified. Both logical and physical security must be verified at each site. Emergency services are needed most when the network is under attack. Cybersecurity is a major topic for all enterprises and service providers especially those that are supporting emergency services. To that end, the standards for next-generation networks include the BCF, which serves as a firewall for internal and external intrusions as well as against application layer attacks. It is not enough to install fully duplicated BCF functional elements. Settings must match the size of the network, and every external and internal caller must be routed at one or more points through the BCF. Before any NG9-1-1 network is turned up for service, it is important to have preprogrammed testing efforts to attack the network from every vantage point and see if it can withstand the attacks. This should be done before live traffic is placed on the network because at that point the exercise has the potential for real impact on service.

Universities, service providers, and vendors have software-based load-generation programs that can help find the weaknesses and send relevant reports to the persons in charge. While standards apply in general, no two networks are identical. Security solutions to meet those standards must be customized to the environment and the parties involved. NENA has a full set of security compliance documents.[14]

Some enterprises that provide their members with telephone devices prefer to send emergency calls that these devices generate to their own police force and security departments for immediate dispatch when they dial 9-1-1. One reason for this is that these organizations have access to floor plans, room information, occupants, and major events that are taking place at a campus or in an organization where they want to be able to direct their internal and/or external police forces to manage emergency situations, including evacuations when needed. This type of internal

14. NENA Security for Next-Generation 9-1-1 Standard (NG-SEC) NENA 75-001, Version 1, February 6, 2010, NENA NG-SEC Audit Checklist NENA 75-502, Version 1, December 14, 2011.

routing of emergency calls must be examined however to ensure that it meets legal, policy, and, most importantly, safety requirements. Where do the 9-1-1 calls from an enterprise route? Does that organization have the ability to immediately contact external forces as needed for additional assistance? Will there be an ever-present operator to answer emergency calls? Is the login and password for the operator(s) controlled and protected? Are the operators trained to handle emergency calls? Will the operator be able to bridge calls to the appropriate PSAP? With the advent of NG9-1-1 there is additional data that can be specifically loaded into the location and emergency services databases in a structured format that allows the enterprise and its security department to show where alarms originate as well as other pertinent information available on spreadsheets, in notes, and in the knowledge of experienced local officers and trained personnel who have been on-site for years.

6.8.1 Enterprises: Opportunities and Challenges

Business campuses usually have Wi-Fi and/or Bluetooth beacons located across their architecture. Hotels and other major enterprises have diagrams and schematics posted in rooms all around their buildings. The NENA standards for i3 provide guidance to use GIS and standard data fields in the new NGCS FEs and in third-party databases, which can appear seamlessly in PSAPs so that any PSAP connected on an ESInet, or connected ESInet to ESInet, can see the information on its online maps and be able to dispatch appropriately. The FCC added Z or vertical-level location requirements in 2020.

Many employees, guests, and students do not have fixed landlines in their buildings and dorms. They rely on the cell phone service and bring portable devices with them. Over 80% of the originating calls in most enterprises are generated from these cell phones or other portable devices.

The OSPs who provide service to an area may have deployed text to 9 1 1 or may not. The students, visiting workers, or guests may already be using text to 9-1-1 in their homes or normal business environments and yet they may not be able to use it when they arrive on an enterprise campus. Moving to NG9-1-1 including text services is valuable in large enterprise settings for proper location of the emergency caller. Cyberattacks can come from ambitious students or employees trying out their IT skills. Students, visitors, or employees may be consuming campus bandwidth to stream videos or use social media apps.

Before entering into agreements and contracts for 9-1-1, meetings are critical to get all the stakeholders involved so that a solid plan emerges. Security concerns are important areas to explore and include so all of the issues are addressed and tested for compliance. Design and operational choices made without considering security may result in extra costs and a less reliable and secure service.

6.9 Reporting and the Enterprise

Commissions have identified various levels of reporting requirements to the OSPs and 9-1-1 service providers that they oversee; none of that reporting directly impacts the enterprise. When an outage or other interruption impacts emergency services, the majority of OSP-and/or 9-1-1 SSP–related incidents are important. In the United

States, outages lasting over 30 minutes and impacting service potential for 30,000 customers or more must be reported to the FCC. The 9-1-1 SSP and OSPs serving 9-1-1 access customers have reporting obligations independent of each other. It is critical to identify the number of people who reside in, or are present in, an area impacted by the outage, so that if and when a problem does arise, the proper notifications are made in a timely fashion. Penalties and other issues, including potential lawsuits, may arise from customers if timely reports are not filed and or loss of life or property occurs during a service impacting outage.

Given the diversity of the OSPs and other challenges in finding customer telephone line counts the government census has been the method for 9-1-1 SSPs to estimate customer impact for reporting purposes. Census data, however, may not reflect the number of people on a campus or served by an enterprise network and so may not reflect the impact of any emergency event on the public. An estimated number of people potentially impacted by location may assist first responders with proper identification and classification of an outage impacting an enterprise. If a PSAP recognizes the total severity of an outage, a mutually agreed upon severity level for problem resolution may be reached. If a business is going through a major change or cutover, the impact of that change should be identified, and changes to persons impacted and severity levels may be in order. Planned maintenance has an impact on the PSAPs, and the 9-1-1 SSPs as well as the enterprise. Contracts need to acknowledge that and have escalation policies, contacts, and strategies for communication 24 by 7 by 365.

6.10 QoS

NENA enterprise documents do not specifically address enterprise connections to OSPs or NG9-1-1 networks. As contracts are written, refer to Chapter 11, which covers QoS standards for the connection agreements between OSPs and NG9-1-1 SSPs. Chapter 9 includes very specific measures the NENA standards suggest for setting up alerts and exception notifications between OSPs and NG9-1-1 SSPs and their subordinate contractors in case they are not providing all of the network connectivity. This may be a helpful starting point for enterprise NG9-1-1 internal monitoring and for use with facilities providers. Enterprises may choose to substitute relevant language to create benchmarks as a guide for contracts. Refer to NENA ESInet Design for NG9-1-1 NENA 08-506, Version 1, December 14, 2011, and augment or edit for enterprise customers. Consider the following unique issues:

- Access to PSAP automatic failover:
    - If diverse access ordered, enterprise facilities shall have automatic failover;
    - The NGCS shall have automatic failover (included in agreement with 9-1-1 SSP);
    - The legacy gateway and security FEs shall have automatic failover (include in agreement with 9-1-1 SSP).

- Minimum bandwidth provisioned:
  - This value is dependent on the size of the network footprint and the volume of calls predicted.
  - 10 megabits/second is generally the lowest amount able to be purchased.

Implementers should document the purchased bandwidth and ensure that there are no blockages (or at least no more than one in 100 calls) before the emergency calls reach the destination. Destination here is the point at which the enterprise can reach for connectivity. It is necessary to have regular meetings to discuss growth plans and expansions and to arrange for changes in purchased bandwidth accordingly. Agree on the priorities and metrics that matter. There are metrics related to location services not included here. Determine how location database information will be shared, how soon it must be loaded after a new service is or turned up for service for a particular line and user, and how errors will be managed from fall out. Agree to accuracy parameters and a means of knowing how the database is performing.

6.11 Location Services

Location services for enterprises and all 9-1-1 user groups are a significant challenge today and with E9-1-1 and NG9-1-1. The FCC has been explicit about the requirements to improve location accuracy. The NG9-1-1 APCO and NENA standards provide significant guidance and requirements support. Chapters 3 and 10 cover this topic. In particular, Chapter 10 describes emerging techniques being used to provide indoor location accuracy.

6.11.1 Location Services in the News

In 2018, new initiatives were launched to find innovative solutions for location accuracy. At the national NENA conference in Nashville in June 2018, Apple and RapidSOS announced a partnership to provide a service that works for E9-1-1 PSAPs and i3 PSAPs and adheres to the parameters of the NG9-1-1 network. Shortly thereafter many other 9-1-1 vendors entered into agreements with RapidSOS. In 2019, Avaya entered into a similar agreement related to its enterprise customers that Avaya itself describes as an OTT solution. The full list of partners is growing and can be found at https://rapidsos.com/ng911clearinghouse/. RapidSOS features include availability to all PSAPs, fast and accurate location from iPhones and Androids, integration with existing PSAP software, and automatic data display. Rapid SOS is committed to remain independent of map display providers; they plan to receive information from the NG9-1-1 clearinghouse and use a secure browser connection requiring no new PSAP hardware or software.[15] Map vendors

15. Mark Fletcher, ENP, Personal Communications, February 7, 2019. Discussion of NG9-1-1 and Enterprise connectivity and location services.

are not a part of the NGCS standards. PSAPs and their 9-1-1 SSPs work to integrate NG9-1-1 capabilities into the map software and hardware they agree upon.The Rapid-SOS website shows: (1) calls coming from user devices calling 9-1-1, (2) calls connected to the NG9-1-1 clearinghouse which (3) researches data already available (4) displays supplemental data, preloaded by the user community, (5) to the PSAP call taker (6) alongside the ANI information, (7) enabling the PSAP to interact intelligently with the caller and 8) the dispatchers to first responders. A key component of the solution uses multiple additional data repositories (ADRs). Chapter 10 details forward-looking location services.

6.11.2 Location Services Standards

The first part of location services is to send the caller's ANI information from the enterprises to the NG9-1-1 network. [To review, there are basic access trunking and signaling alternatives described in NENA 03-502, April 11, 2003 (Original).] The list below is the typical Enterprise signaling access to the NG9-1-1 network.

- Dedicated CAMA/MF trunks from the PBX to the NGCS via the LNG;
- Dedicated CAMA/MF trunks from the PBX to the serving OSP CO where calls are switched and transported to the legacy E9-1-1 SR that sends the calls to the NGCS via the LNG;
- ISDN-PRI trunks configured to pass "9-1-1 valid" caller location identification (CLID) to the serving CO where calls are switched and transported to the NGCS via the LNG;
- ISDN-PRI trunks configured to pass "9-1-1 valid" CLID to the NGCS via the LNG.

The same NENA document states that, "all station lines associated with the PBX, including off-premises extensions and lines served by sub-tending remote switch modules are located within the service area of the 9-1-1 selective router that serves the location of the main PBX. In cases where there are station lines located outside the area served by the 9-1-1 SR that serves the location of the main PBX, a configuration using multiple CAMA/MF-type circuits connecting the PBX to more than one 9-1-1 selective router may be feasible if the PBX can be programmed to accurately route 9-1-1 calls to different outgoing trunk groups"[16]The standard should be reviewed in its entirety if there are remaining CAMA/MF trunks in the network. In addition, the NENA standards caution the reader about testing, as follows: "THE NEED TO TEST THE ABILITY TO PLACE 9-1-1 CALLS FROM EVERY PBX THAT USES ISDN-PRI TRUNKS CANNOT BE OVER EMPHASIZED."

16. NG9-1-1 SSPs may cover multiple states or 9-1-1 jurisdictions. This can include areas converted to NG9-1-1 and those remaining on legacy SR connectivity. Network planning, engineering, testing, and end-to-end call completion must be documented thoroughly. Diagrams should be made available to the persons doing ongoing operations and troubleshooting with contacts in case of call failures after initial setup.

6.11.3 9-1-1 Test Call Ground Rules

One of the most challenging parts of the 9-1-1 implementation process is obtaining the test numbers to use for the pretesting[17] and cutover testing phases of the conversion to NG9-1-1 project. During contract negotiations with enterprises and OSPs, it is essential to agree that originating test numbers are shared with all parties. In the case of OSPs, a sample of lines is generally accepted for successful testing by commissions and PSAPs. The responsibilities for what entity provides the test lines in the case of the enterprise is between the enterprise and the database managers of the NG9-1-1 SSP and the PSAP managers who have access to prior test call numbers and access to individuals in the community who may agree to be a part of critical testing. The information that follows refers mainly to PBXs. Centrex callers are subject to many of the same rules and processes except where noted.

Enterprise test numbers must be preloaded into the NGCS databases along with all GIS and relevant address information in the same format as the other lines in service. The NG9-1-1 databases must be built for each customer before cutover so that sample testing or line station number testing can be done—whichever is appropriate per regulations and agreements. The use of P.O. boxes and rural route address numbers is insufficient to get the CLID information to map into the PSAP. The addresses of the OSP CO test line stations must be added to the DBMS records when the OSP is doing any testing from their location. Cooperative agreements must include testing. The enterprise database records must be reformatted and augmented to meet all of the NENA database standards. Database management and testing can be one of the longest lead times in the migration to NG9-1-1. For many OSPs, there is a requirement to provide test numbers by NPA NXX and line groups to the LERG. That does not include test numbers from enterprises. The database process, responsibilities, challenges, and expenses are some of the reasons for the recent actions to move to OTT solutions as described in Section 6.11.1. In point of fact, any solution requires extensive testing and advanced planning. The OTT vendors provide their own guidelines.

6.11.3.1 Enterprise Test Planning

Assuming that the NG9-1-1 GIS database in the NGCS and the location databases is preloaded, pretested, and ready-to-use, the following rules have been suggested by NENA. Refer to NENA documentation, NENA 06-003 Revised August 2004, Private Switch (PS) E9-1-1 Database Standard.

6.11.3.2 Project Manager

Ensure that all PSAPs are prenotified that testing is scheduled and that they are able to handle the load. An emergency can arise making it impossible to perform the planned testing. For further information, the authors defer to the NENA MLTS testing guidelines.

17. Chapter 7 covers testing for OSPs and Enterprise. Wireshark and or other tools should be used to trace the calls end-to-end and determine the failure location for problem resolution. Monitoring systems such as those described in Chapter 9 should be used by the OSPs, the 9-1-1 SSPs, and any third party involved in the end-to-end call flow.

6.11.4 Enterprise Database Management

Enterprises may send PS/911 ALI records directly to their OSP, DBMS, or third-party DBMS provider. The term DBMS herein refers to both primary and third-party DBMS providers. Third-party DBMS providers perform data preprocessing, validation, and management services. The third-party DBMS provider serves as an interface to the primary DBMS provider. It is the responsibility of the enterprise to establish relationships and any agreements required for the provision of PS/911 data. The enterprises must coordinate PS/911 database and network requirements with their 9-1-1 SSP and the DBMS provider to ensure the proper operation.

6.11.4.1 Updates

Changes adds and deletes involve a significant workload for the DBMS team. GIS information must constantly be updated by the PSAP GIS coordination team or the 9-1-1 SSP, whichever assumes that responsibility so that the routing of the calls to the correct PSAP is assured. Newer location identification methods are under discussion in standards bodies and by entrepreneurs and vendors offering low-cost or free apps to assist with automation of the caller's location. Refer to Chapter 10.

6.11.5 Enterprise Database Security

NENA-06-003 Revised August 2004, NENA Private Switch (PS) E-9-1-1 Database document say, "Security is an inherent component of data transfer and is necessary to provide assurance of the confidentiality and integrity of the PS/911 ALI records. It could be considered a legal requirement, not only for privacy considerations, but in addition, for accuracy of the information transferred. Collection, creation, manipulation, storage, retrieval, display, and transmission of PS/911 ALI records expose that information to modification or destruction. The reliability of hardware, software, communications, application and human factors shall be considerations in the selection, design, and implementation of any system for data preparation and transfer. Security controls such as encryption, redundancy, and password levels are some features that shall be considered in the selection of any system or method used for PS/911 ALI data preparation and transfer. Realistic methods of minimizing or eliminating risks shall be the responsibility of the Enterprise and the DBMS provider."

The NENA-06-003 Revised August 2004, NENA Private Switch (PS) E-9-1-1 Database document includes a significant amount of detailed information that must be used when implementing enterprise databases.

6.11.6 Database Summary

"Telephone line stations or inaccurate location information will be displayed to the 9-1-1 Call Taker and such inaccuracies may delay the response of emergency services." (https://www.fcc.gov/general/9-1-1-and-e9-1-1-services)

Where the Enterprise does not provide true PS/911 functionality and the MLTS system is configured to send the main or billing telephone number as ANI when a 911 call is placed, then the main or billing address associated with that number will display on the PSAP screen. This may not be the caller's actual location; however, it will aid in locating the caller.

Should the Enterprise or OSP or Third-Party enable station level ANI without providing PS/911 ALI records, 9-1-1 calls placed from within the MLTS system will follow the established Default call route and may not be answered by the correct PSAP. It is highly recommended that Enterprises provide true PS/911 functionality. True PS/911 service may only be achieved by accurately maintaining timely and accurate MLTS telephone line station records in the 9-1-1 DBMS.

Where an MLTS system is used to provide telephone service over a large area or campus environment, special consideration shall be given to selecting an alternate callback number in the event the caller is incapacitated or has left the area. In these cases this call back number may be a vital link to the facility where the emergency has occurred."

6.12 Location Identification

"With the advent of VoIP and Wireless phones, a lot of discussion has been devoted to locating emergency callers. Critically relevant is the large numbers of individuals who on a daily basis might use their office telephone or other device in an Enterprise MLTS environment to dial for help.

Identifying the location of individual callers during an emergency is a challenge that involves the individual, the organization, the governmental organizations responsible for providing public safety services and other Third-Party entities including those delegated various responsibilities by the government. The trend is significant. In addition to private residences where a rapid growth in the use of VoIP is occurring, Enterprise telecommunication systems are seeing multiplicative growth in the use of similar technologies."[18]

6.12.1 Location Services for NG9-1-1

There are many of the new terms and acronyms in use for NG91-1 location services as they apply to the DBMS function and processes. Database management users must follow the NENA standards and the various vendor product specifications to prepare for NG9-1-1 and implement it successfully. There are no shortcuts. Needless to say, early adopters of NG9-1-1 have embraced this responsibility to provide accurate database information for NG9-1-1 systems. There is no published reference about enterprise NG9-1-1 location success. When this book is updated, the reality of NG9-1-1 database management for the enterprise will be included.

6.12.2 Location Services and Providers

As NG9-1-1 with i3PSAPS has been deployed, the diversely routed duplex e2 datalinks with ALI database providers were moved from the individual E9-1-1 PSAPs

18. NENA 06-502, Version 1, October 25, 2008, Industry Common Mechanisms for MLTS E9-1-1 Caller Location Discovery and Reporting Technical Information Documents (TID) and NENA 08-752 Issue 1, December 21, 2006, NENA Technical Requirements for Location.

to the redundant NGCS in each data center or CO hosting the NGCS. Figures 6.1 and 6.2 have high-level diagrams that include the e2 datalinks. It is important to note that E9-1-1 PSAPs have generally chosen to interface to only one external third-party database provider. If that provider does not service all of the OSPs needing external database services, then there are crossover agreements between the third-party providers to obtain and return the location address and callback number to the E9-1-1 PSAP directly. With NG9-1-1 and the move to the data centers or COs, the e2 datalinks are usually carrying significantly more traffic, and the NGCS needs direct duplex diverse connectivity to each of the major database providers. Those providers until recently were West Public Safety, Comtech TCS, and Bandwidth. With the entry of RapidSOS to the marketplace, new links or connections that meet their specifications must be established in a redundant fashion for real-time database access either to the PSAP if it is still an E9-1-1 PSAP or to the NGCS if the PSAPs have been converted to i3 PSAPs. Care should be given where virtual private network (VPN) connections are used. The authors have seen intermittent VPN failures related to database connections. The use of default routing and default PSAPs is mandatory in the event of such failures, but the dispatch to first responders may be delayed if the PSAP answering needs time to locate the caller and or even determine the correct PSAP to handle the calls.

Enterprises with VoIP OSPs require third-party ALI database systems to provide the location information. Enterprises connected to or served by circuit-switched or TDM OSP-owned technology may find that the OSP database has migrated from a traditional ALI-MSAG database provider to the diverse, redundant NGCS housing the NG9-1-1 routing databases. An NGCS FE called the LIS is the interface for loading the new GIS databases so the ECRF can query them when the calls arrive. It is theoretically no longer essential to rely on the former third-party providers that managed the external E9-1-1 wireline ALI databases. For enterprises these options require business decisions. The interfaces to the LIS are new to OSPs and the enterprises and/or their third-party agents that assist in the database and 9-1-1 process. Third parties, including RedSky and Motorola, or OSPs with a database business model and systems, especially for their Centrex customers, may be involved. Open communications and sharing of options are essential.

6.12.3 Location Services Definitions

Download the Glossary of Terms and Acronyms for this Handbook. The terms are essential to the NG9-1-1. Refer to NENA 08-752 Issue 1, December 21, 2006, NENA Technical Requirements for Location Information to Support IP-Based Emergency Services

6.13 Summary

Database management and location services are among the most critical responsibilities associated with the enterprise's successful transition to NG9-1-1. Building a successful NG9-1-1 infrastructure is the focus Chapters 1–5. Chapter 2 addresses business issues and steps in choosing the 9-1-1 SSP and the systems integrator and project managers to manage the NG9-1-1 transition. Experienced GIS and database management leadership is critical from the beginning of the planning process. The building of the database and selection of the vendor solutions and design is the first step toward developing a timeline for the project. From the perspective of the NG9-1-1 system provider, evaluation of the various options requires someone who understands the enterprises in the footprint to be served, how they are connected today, and the future opportunities to migrate those customers to NG9-1-1. Neither the project timeline nor the budgets can be established without identification of and understanding of the enterprises in the footprint. Legal and regulatory matters mean that access to legal and regulatory personnel in the area to be migrated are essential. Refer to Chapters 2 and 5 for additional concerns that are important to any 9-1-1 project. Chapters 7, 8, 11, and 12 are essential when dealing with enterprises in the context of the total project management lifecycle and budget.

Selected Bibliography

NENA 03-502, April 11, 2003 (Original) Trunking for Private Switch 9-1-1 Service.

NENA-06-003 Revised August 2004 Private Switch (PS) E-9-1-1 Database Standard.

NENA 06-502, Version 1, October 25, 2008, Industry Common Mechanisms for MLTS E9-1-1 Caller Location Discovery and Reporting Technical Information Documents (TID).

NENA 08-752 Issue 1, December 21, 2006, NENA Technical Requirements for Location Information to Support IP-Based Emergency Services NENA 08-752 Issue 1.

NENA 75-001, Version 1, February 6, 2010, NENA Security for Next-Generation 9-1-1 Standard (NG-SEC).

NENA NG-SEC Audit Checklist NENA 75-502, Version 1, December 14, 2011.

NENA-ADM-000.22-2018, 04/13/2018. NENA Master Glossary of 9-1-1 Terminology.

CHAPTER 7

Test Plans

Things will go wrong in any given situation, if you give them a chance
—Murphy's law

7.1 Introduction

When deploying an IP-based NG9-1-1 communication system, testing is the project phase that determines the suitability of the system, the timing for cutover, and ultimately the success of the project. Each deployment of NG9-1-1 starts with a set of formal requirements such as the RFP and the most current version of NENA standards. The desired results include verification of requirements, including functionality, redundancy, resilience, diversity, and scalability. A complete testing plan is essential, so that all parties know when each location is ready for cutover and service.

The requirements for the NG9-1-1 infrastructure, network, and access are defined when the RFP is prepared. The RFP delineates the standards as they are known and available at that time, as are the services and unique attributes of the communities to be served. A first step when developing the test plans is to review the RFP and break it into logical areas to be tested for compliance. Elements of NG9-1-1 testing include: the infrastructure and database (Section 7.2), the network (Section 7.3), access (Section 7.4), and end-to-end (Section 7.5). All of these tests must be completed successfully before cutover (Chapter 8). Depending on the cutover strategy chosen, the cutover test plan is a subset of the end-to-end test plan. Note that before doing any tests, it is important to have all IP addresses, the DNS, SS7 PCs, diagrams, NPA NXX lists, LATA configurations, agency boundaries, neighboring agency borders, and CLLI information, documented and available. Be prepared for all contingencies and have all legal and regulatory matters in progress or completed. As the project team moves forward there are a few things to remember: The test plans state clearly what is being tested, what the expected outcome should be, how success is measured, and the results, pass or fail. A process needs to be in place to create trouble tickets, to track tests that fail, to create a history of tickets that have been resolved, and to ensure closure. In some test case failures, the hardware or software may not be operating per specifications and an escalation may need to be made to clarify or enhance standards for the rest of the world to benefit. The project manager tracks progress; the lead technical team experts conduct the testing.

7.2 Infrastructure

Sometimes the hurdles aren't really hurdles at all. They're welcome challenges, tests.

—Paul Walker

Preparation: By the time testing is scheduled, the project managers have ordered and installed the ancillary equipment and interfaces required at each PSAP and data center location. The ESInet and core are part of the network test plan and tested along with connectivity to each PSAP's equipment and functionality. After the network is tested, access and connectivity follows. Service is fully addressed when end-to end testing occurs. Infrastructure and database testing includes environmental and physical security testing, which should be completed ahead of the installation of the NG9-1-1 equipment. For PSAPs, this may include, but not be limited to local switches and routers, cabling, furniture, computers, monitors, power, and conditioned space in the call-taker area and racks and power in the IT area to accommodate all phases of the test plans. Data centers must have completed legal security agreements and the necessary racks, mounting boards, and virtual or physical cage spaces readied, power, and grounding, fire suppression, and environmental requirements must be met. The NGCS and ancillary equipment should be installed and ready to test. The NOC must be ready for training, tools, monitoring and ticketing systems, and skills. Conference bridges are essential for any test plan; multiple bridges may be required. Project managers and lead teams determine the test plan leader, the documentation manager, and the time and details for each plan to move forward. Each test plan defines those who directly participate and those who must be available on call. It is suggested that the team use a spreadsheet like Excel, throughout the testing process; this book does not include spreadsheets. Expertise in managing large spreadsheets with significant details is critical. This applies to testing and through to the cutover plans.

7.2.1 Alpha/Beta Testing

A significant amount of testing of NG9-1-1 components and their integration with all other components of the NG9-1-1 network have been completed ahead of what is traditionally called alpha and beta laboratory testing being deployed to any customer site for full systems integration. Prior to any site deployment testing, industry standards testing is complete to ensure that the products from any vendor interoperate with products from any other vendor. There are stages to alpha and beta testing. Vendors have their own laboratories where alpha or first testing takes place. Some vendors choose to develop a full suite of products and have them pre-integrated in their own labs long before bringing the products to beta testing or integration testing with other vendors, and before bringing them to the open market. Other vendors may only have one basic type of FE or a few FEs such as the systems for database management (i.e., LVF/LIS), logging and recording, mapping, repair, and monitoring. The 9-1-1 SSP and systems integrator chose the suite of products to be deployed from their experience with the vendors and have been trained on the integration of these products. Before the 9-1-1 SSP bids on the project, they typically have an equip, furnish and install (EF&I) contract arrangement with the vendors they represent and have an understanding of the costs and timeline to deploy those products along with the manpower and tools to test them once they are introduced into the customer sites.

7.2.2 Integration and Conformance Testing

NENA sponsors vendor integration processes to test the interoperability of products between vendors against the NENA standards. Additionally, NENA engages various university laboratories to serve in the integration testing process as neutral parties. Universities in the United States have included Texas A&M and the Illinois Institute of Technology. Each testing event uses NENA standards for the particular area under test. As this book goes to press, not all standards are in final form or have been through NENA testing. Therefore, first deployments across the country have many parts of the network going from beta testing right into field deployment testing.

Per NENA.org, "NENA is dedicated to ensuring that the goals of NG9-1-1, both technical and non-technical, are met. NENA views a range of testing programs as a critical component to accomplishing standards based NG9-1-1. NENA understands that it is the vendors of NG9-1-1 elements that ultimately deliver the interoperability NG9-1-1 promises. Therefore, NG9-1-1 testing events sponsored by NENA are aimed at supporting the vendors in their testing with each other. These Industry Collaboration Events (ICE) bring together vendors in an open, supportive, and collaborative environment that foster a spirit of technical cooperation. Taking part in or success in testing at an ICE event does not confer any formal NENA certification for a vendor's products"

"While NENA has played a central role in the creation of the ICE program, it is our intent to include all stakeholder groups and we are open to partnering with other industry organizations in the creation and implementation of this program. This is evidenced by the makeup of the NG9-1-1 ICE Steering Committee, which has seats representing the vendors, government users and buyers, other industry associations, government organizations, NENA Development Group Leadership, and NENA Senior Staff. The Data Gathering and Reporting Committee is responsible for reporting test results." The ICE steering committee is charged with overseeing all planning and execution of these testing events. Table 7.1 lists past and upcoming NENA ICE testing events.

While NENA has released an RFP for vendor conformance testing, this process has not been completed. As of this book's publication date, no conformance test plan is available.

7.2.3 Prequel Site Testing

As described in Chapter 3, there are many infrastructure requirements: environment, heating and cooling, layers of power, grounding, timing and synchronization,

Table 7.1 NENA ICE Testing

Event	*Dates Held*	*Locations*	*Test Purpose*
ICE 1	Nov. 3–5, 2009	Texas A&M University in College Station, Texas	Test location, routing and delivery of NG9-1-1 calls
ICE 2	May 24–28, 2010	AT&T Center for Learning in Irving, Texas, with network support from Texas A&M University	Test NG9-1-1 transitional elements
ICE 3	Nov. 29–Dec. 3, 2010	Brazos Valley Council of Government (BVCOG) site in Bryan, Texas, with network support from Texas A&M University	Location information
ICE 4	Nov. 14–18, 2011	AT&T Center for Learning in Irving, Texas, with network support from Texas A&M University	Call routing based on-LoST hierarchy
ICE 5	Oct. 15–19, 2012	Illinois Institute of Technology, School of Applied Technology, Real-Time Communications Lab, Wheaton, Illinois	Accessibility to emergency services
ICE 8	Nov. 4–8, 2013	Illinois Institute of Technology, School of Applied Technology, Real-Time Communications Lab, Wheaton, Illinois	Interoperability with recording and logging components
ICE 6	Nov. 9–14, 2014	Illinois Institute of Technology, School of Applied Technology, Real-Time Communications Lab, Wheaton, Illinois	End-to-end test of i3 architecture
ICE 7	Mar. 19–22, 2017	Illinois Institute of Technology, School of Applied Technology, Real-Time Communications Lab, Wheaton, Illinois	Apps, EIDD, additional data, GIS/mapping
ICE 9	Week of 2/22/2021	Illinois Institute of Technology, School of Applied Technology,- Real-Time Communications Lab, Wheaton, Illinois	Global interoperability of next generation emergency calling using Internet multimedia

"Each ICE has a focus for the testing to be done" The ICE Steering Committee has the overall rules for readiness for ICE testing. "Government organizations and their consultants are encouraged to support this program by adding questions related to participation in the events to any RFI or RFP they develop. Boilerplate for this has been developed and is in use by multiple jurisdictions. Recently, the National Association of State 9-1-1 Administrators (NASNA) has endorsed this effort and encourages all vendors to participate." Refer to NENA.org.

and physical and logical security. Refer to Table 7.2 and verify that every aspect has been tested, and that alarms are going to the monitoring system(s), the NOC, the project manager(s), and any vendors and/or subcontractors. Ensure that these alarms are included in these tests and results documented. The overall project manager should have dashboard milestones for every step of this process. Before testing begins, the network monitoring systems should be installed and ready to capture alarms and provide alerts; refer to Chapter 9. Additionally, ensure that the equipment in the data center, along the ESInet, and in each PSAP and database location is integrated with the monitoring system and set to alert at appropriate threshold levels. Breaking a threshold should trigger alarms and alerts for each critical item, with alerts sent to the monitoring system.

The 9-1-1 SSP and SI must have software and hardware tools to do the testing or have a plan to gain access to those test tools and the personnel qualified to facilitate testing at every level of the project. Tools may include, but are not limited to protocol analyzers for IP and SS7 systems and free software tools such as Wireshark to initiate and trace the calls and queries through the network, verify they are

Table 7.2 Infrastructure Tests, Thresholds, and Alerts

Sample Checklist

Item Alert Set ranges to vendor specs	*ESInet POIs 1-n*	*Data Center 1*	*Data Center 2*	*PSAP 1-n*	*Alarm to OSS Pass Fail–Date*	*Monitoring Alert sent to NOC Pass Fail–Date*
Heating:						
Top range:						
Low range:						
Cooling:						
Top range:						
Low range:						
Building ground Verified						
Power 1 Commercial						
Power 2 UPS						
Power 3 Generator						
Power 4 Fuel supply						
Power 5 Battery fluid levels						
Power Contact names and numbers list						
Physical security—itemize per locations Ex (1) doors (2) windows (3) human security—look for protocols and test alarms						
Logical security per OSS or system.						
List and show levels of alerts and alarms. Check protocols per site. Include password protection protocols						
DNS server						
IP PBX						

Create the worksheet for these verifications and tests and maintain information in a secure location. Contacts for each location are critical and define the responsible parties. Names of vendors and 24 by 7 reach information must be included. Escalation procedures must show who is next to be alerted and or notified if the primary contacts are not available. The 9-1-1 SSP is ultimately responsible unless otherwise noted.

completing, and see where they are failing. In some of the test cases, failures are the correct answer. How the calls fail is important.[1]

7.3 Network Testing

To set the stage for testing the project, the 9-1-1 SSP and the SI have chosen the ESInet provider(s), the data center locations for each NGCS component, and ancillary

1. Example of good test failure: All Trunks to a PSAP are made busy and the calls are sent to the alternate PSAP where all Trunks are also Made Busy. Calls receive the correct response of 120 IPM, fast busy, or an announcement to originate the call again. The test plan will include expected result and actual result will be noted.

equipment to complete the transition to NG9-1-1 1. The SS7 provider has been chosen for the legacy SS7 signaling. The cybersecurity components were purchased as well as all legacy gateway interfaces. Each PSAP location is fully identified, and decisions about upgrades to i3PSAPs or using legacy E9-1-1 PSAPs have been made. A project plan has been developed so that all parties know when each location needs to be tested and turned up for service as well as the parties required for the tests. Note that if the area to be converted has integrated text to 9-1-1 in a pre-NG9-1-1 environment, as encouraged by NENA, and the FCC for Americans with Disabilities, those tests must be added to the test plans. If text to 9-1-1 is being introduced at the same time as the conversion to NG9-1-1, testing that new service must be included in all the test plans. Augment the plans to cover all unique circumstances for the area. Chapter 10 addresses future capabilities.

At this time, the SI has delivered products that were pretested in laboratories and NENA ICE events to each site. The 9-1-1 SSP and/or the project leader for the agencies have prepared each of the receiving locations and purchased and installed the equipment from their checklists. If there are ancillary systems such as MIS or OA&M systems, those items have been purchased and installed. Racks and mounting space and cages and mounting boards are ready for connectivity to the SS7 providers, OSPs, text-to-9-1-1 service providers, enterprises, and the PSAPs.

The network test plan is essential at this point. Tests follow the network plan, testing the products purchased and old technology, such as CADs, or map systems, to be retained for the NG9-1-1 project. Chapter 3 provides a generic name for each of the NGCS components. The vendor product names may not match the functional names, and in some cases a particular NGCS FE hardware component may handle multiple functions depending on the way that vendor chose to interpret the standards. This section describes each system for the convenience of the project manager and testing team. There may be entirely different teams doing tests for various parts of the network, and then once they come together, integration and interoperability testing is required until every aspect of the network testing is complete. Always refer to NENA requirements if the RFP is not specific in terms of performance.

7.3.1 ESInet

The ESInet is most likely made up of purchased bandwidth on a shared system, especially in regional NG9-1-1 projects. When it comes to the network, document the bandwidth for each piece of the ESInet and that required for one data center to completely fail over to the other with extra capacity for high traffic load. Refer to Chapter 3. Use the design engineer's bandwidth and capacity specifications to test. Whether a cable or fiber is cut, or an electronic interface device or router along the ESInet fails, the engineered solution should allow calls to move in another direction automatically to complete to the proper destination. Part of testing the functionality includes assurance that the proper IP addresses have been assigned. A list of the design engineer's IP addresses per FE and its alternate across the whole architecture is essential.

7.3.2 Data Centers or Central Offices

The duplicated data centers or telco wire center locations must be fully built out, ready for network testing. Vendor installations must be completed with the vendors doing their initial failover testing before turning the FEs over to the project manager(s) and testing team. If the data centers are owned by third-party providers, their technical NOC and testing team and supporting parties must be ready for the network tests to commence. At this point building failures must be simulated to determine that each alternate data center can see the failure and respond as required. The alternate data center(s) must be prepared to handle all the traffic load. If the data centers are owned and operated by different parties, they may each have their own NOCs, and they may each schedule their own maintenance with their clients. In all cases, there must be assurances that these organizations notify the 9-1-1 SSP of any planned activity and be prepared to schedule important activities only after determining that the other data center(s) do not have the possibility of a disruption at the same time. Include tests to see that all NOCs can receive exceptions appropriate to their areas of responsibility. The test plan can be annotated to include the responsible NOC where multiple organizations are involved.

7.3.3 NGCS

The vendors and SI test each FE as it is installed and the interoperability of each FE after it is installed to each data center. The technical lead from the 9-1-1 SSP project team is a partner in this testing at each stage of the process. Once it is turned over to the 9-1-1 SSP, additional tests are performed according to the test plan developed from RFP requirements, consistent with the design and engineering plans of the regional or statewide entity that made the NG9-1-1 system purchase. Settings for routing and alternate routing, database parameters, and all unique aspects of the build to the particular client must be addressed. Database testing (i.e., GIS/LVF/LIS) takes place well ahead of the network testing and continues ongoing throughout the project and beyond. The database vendors and the database and GIS teams remain active and available throughout NGCS testing. The LIS is tested as part of the database testing module along with SO interfaces. Section 3.6 addresses testing databases. The LSRG is also critical to the access testing process. The physical gateways get tested for the full capabilities when calls enter the network. The fact they are present and can be failed over automatically to the alternate data center is critical for the network test plan.

7.3.4 PSAP

Once the testing is completed at each data center, the first PSAPs to be cut over must be tested with the NGCS in each data center to verify connectivity, reliability, and failover as specified in agreements. PSAP testing includes testing ESInet interfaces to each PSAP. PSAP testing includes full integration of the i3PSAP with the other local systems and processes. A process may include, but not be limited to: logging in and out of the PSAP, seeing if the performance demonstrated by the vendors operates smoothly, transfers of calls to participating and adjacent agencies, default

PSAPs receive the calls they should and can record and transfer them to all of the agencies they serve, language services transfers work, telematics interfaces operate smoothly, and if all normal operations operate as required. The PSAPs as users of the NG9-1-1 network must be fully tested before any live traffic can flow to each workstation. Test full integration with the local phone systems and other products such as the ACDs, CAD, MAPs, and all existing alerts and alarms that exist for the normal function of the PSAP.[2] Figures 3.1–3.4 in Chapter 3 are useful for reference. Project-specific diagrams should be developed to show the full configuration with the switches, routers, and POIs to the ESInet and how everything is designed to operate. Think of this as the blueprint of the network for each location and the call flow. Preceding Chapters 1–6 describe the process of building the necessary material and references to have available. Testing depends on knowing the full scope of each location. If there are legacy PSAPs in the architecture, the LPG testing is included here. There are likely no exactly duplicate PSAP locations. Each PSAP has been accustomed to having its own equipment; sharing these assets with the region or state is a new experience. The role of the default PSAP and the alternate default PSAP must be addressed in the network testing for PSAPs. Unique tests, including transfers, must be addressed. In preparation, operations processes have been agreed upon, deciding who responds when there is system trouble and how to work as a team. In effect, the plan is testing new or modified standards and processes at the same time as the functionality of the equipment. Trouble ticketing, alerting, and alarming may be impacted: The test plan needs to state: (1) what is being testing, (2) what the expected outcome is, and (3) how success is measured and, subsequently, document the results as pass or fail. The process includes trouble ticket creation. In some cases, the products may not be operating per standards and an escalation may need to be made to clarify or enhance standards for the rest of the world to benefit. Here the project manager tracks progress, with the lead technical team experts as identified in Chapter 2 in charge. (See Section 7.3.7.) The local PSAPs should not be burdened at this point in the testing process. The agency project manager and the lead team from the agency need to be aware and help to make key decisions when there is a jeopardy.

7.3.5 SS7 Integration

Even though it is the 21st century, the SS7 technology from 1975 makes up the signaling for a significant number of calls (a vast majority for the foreseeable future). Almost all OSPs still use SS7 for signaling. In the network test plans, it is essential to test the newly installed A links with the 9-1-1 SSP's SS7 provider. That integration comes to the NGCS through legacy gateways. Legacy gateways vary by the installation. They are meant to be interim interfaces between the IP networks and the legacy world outside the i3 NG9-1-1 network. Even if the work has been outsourced to manage SS7 integration, it absolutely must be tested in the network

2. Integration with each PSAP's local environment requires an inventory of what is onsite, validation that each capability is needed going forward, is integrated and proven operational when the process concludes. Local personnel accept a project based on their ability to operate seamlessly through the upgrade to NG9-1-1 and afterwards. PSAP standards and processes must be seamless. A PSAP answering a call must be able to provide quality service to all callers.

part of the test plan using third-party SS7 monitoring systems and NOC interfaces. Determine who needs to be on the SS7 integration team. SLAs should be in place with third-party providers. Include the SS7 and all legacy types of interface testing in your detailed test plans. Full SS7 testing is only possible when the OSP test plans are prepared and utilized. In our experience with test plans, SS7 flow diagrams were used to assist the team in their testing efforts.

7.3.6 Security

Test plans for the network are extensive and time-consuming. The good thing is that no matter how many OSPs and PSAPs are connected to the ESInet and the duplicated data centers, once the bulk of the network testing is completed, much of it does not have to be repeated. Cybersecurity guidelines from NENA outline the required security testing. ICE has yet to host the event on cybersecurity. (Table 7.1 lists the ICE testing schedule.) Logical and physical security testing is part of the infrastructure plan. The network test plan is different and discussed in Section 7.3.7.

7.3.6.1 Overload and Stress Tests

It is essential to plan and complete stress and overload testing, which includes DDoS, before the cutover timeframes are determined. Some SIs may have their own overload and stress programs. In other cases, university laboratories are ideal places for contracting that testing. The testing source must be able to generate traffic loads and overloads of a significant volume. The testers need access to the internal FEs unfettered by the BCF firewalls since it is possible for load and overload to be generated by the PSAP personnel themselves either inadvertently or not. It is important to stress test the BCFs from every possible perspective. The authors have observed important findings from these types of tests resulting in adjusting limits and finding places where the BCF was set for too low a limit for call volumes. If the network is only testing a single call through a FE at a time, it may appear to operate properly. Adding loads simulates performance demands in busy hours, busy seasons, and times of local or national disasters. Chapter 10 addresses the potential role of universities in this process.

7.3.7 Sample Network Test Plan

Below is a basic template to use as a start point for a Network Test Plan.

7.3.7.1 Overview

Each test case has been identified with a unique test case number, test case description, expected successful outcome of the test case, and actual outcome of the test case. There may be many test cases for which the explanation of the expected outcome is captured in the text of the test case description, and no additional notes are included. These are to be shown as a blank "expected outcome" field.

7.3.7.2 Test Plan Limits

There are some tests that do not require any traffic or call attempt results, and those can be performed any time. There is a distinction between access testing and network testing of the NG9-1-1 network. Refer to Section 7.4 for sample access. Access testing involves direct participation by the OSPs and large enterprises. Telematics providers, and anyone with any device that may be accessing the particular NG9-1-1 network under test, today or at cutover, must be involved. Refer to Chapters 5 and 6.

7.3.7.3 Preservice Testing

Network test calls will be made prior to live activation of the 9-1-1 service and access test calls will be made to validate end-to-end performance. Some tests are designed to demonstrate performance of the network.

7.3.7.4 Network Connectivity

Section n of the actual network test plan describes the tests performed to validate NG 9-1-1 network connectivity, logging, monitoring, and administration capabilities.

7.3.7.5 Load Testing

Section n describes the tests performed to determine that key components of the network design functioned in a resilient and secure manner and performed well under a simulated load. Section n describes the tests to be performed and documented.

7.3.7.6 Vendor Testing

Section n describes the testing performed with various vendors of the FEs, taking advantage of tools and automated procedures used during installation validation.

7.3.7.7 Sample Test Plan Form Types for all testing

Below are the various forms necessary to create a Test Plan.

1. Standards trouble reporting and resolution template;
2. General trouble reporting and resolution template;
3. End-to-end testing template;
4. Cutover testing template derived from end-to-end test plan.

7.4 Access Testing—OSP and Enterprise

The access test plan is developed to test the functionality of the NG9-1-1 calling system being deployed by the 9-1-1 SSP for a public safety agency, or a member of a regional project. The objective of this testing is to ensure that all OSPs and directly connected enterprise PS/ALI switch providers have successfully transitioned their 9-1-1 traffic to the new system, including turning up all appropriate physical and logical connections, and provisioning equipment in their networks to POIs with the NG9-1-1 system. The testing is intended to include all OSPs in the footprint, including wireline, wireless, VoIP, and directly connected enterprise PS/ALI customers. In the remainder of the document, reference to OSPs includes those enterprise PS/ALI customers unless designated otherwise.

The OSP test plans define the test procedures for acceptance testing between the OSPs' 9-1-1 access networks and the NG9-1-1 NGCS. All testing shall be coordinated by the 9-1-1 SSP, in cooperation with each participating OSP or enterprise PS/ALI.

Test calls are placed by the OSPs, and the full test team confirms that the correct ANI and ALI are received and verifies that the OSP networks deliver 9-1-1 calls via SS7, ISDN PRI, MF, or SIP protocol to the correct NG9-1-1 data centers and the NG9-1-1 NCGS. The 9-1-1 SSP and the OSPs work together to resolve any integration issues that arise during testing. This testing will include on-net transfers and the routing and/or return of split exchange traffic with LECs in areas where split exchanges exist. This test plan will be used in conjunction with other test plans developed for the project. These OSP tests are not performed with live PSAPs. The PSAPs are still in test mode in the backrooms or IT space of the PSAP that will be eventually cut over to this configuration. There are several related diagrams and tables that must be referenced. These diagrams and tables will assist with planning and executing the test cases described (Chapter 5, Figures 5.1–5.8). Some of them listed as follows:

1. SS7 ILEC diagram;
2. SS7 ILEC legacy selective router for split exchange transfers diagram;
3. SS7 CLEC diagram;
4. SS7 wireless diagram;
5. SS7 VoIP diagram;
6. PS/ALI diagram;
7. e2 datalink for MPCs and VPCs diagram;
8. SIP diagram;
9. Carrier listing document including name and type of carrier, SS7 PCs, and legacy SRs;
10. IP Addresses for SIP interfaces;
11. NPA NXXs with CLLI per OSP access carrier per location, host/remote and more on a series of Excel spreadsheets.

The key components of a sample OSP test plan include the following:

- Test objectives and guidelines.
- Equipment and software.
- Test cases:
    - Transport;
    - Signaling;
    - Routing and failover;

  - Split exchange;
  - Text to 9-1-1;
  - Site acceptance;
  - SLAs as applicable by OSP or OSP and enterprise.
- OSP, test numbers per wire center, per rate exchange, per class of service.
- Phases for enterprise PS/ALI customer testing as described in the test plans.
- Network diagrams that show which OSPs route through the SRs and the OSPs that will route directly to the NGCS via the BCF or via the LNG.
- The public safety agency must work with 9-1-1 SSP to conduct coordinated testing with the OSPs when any of the following occur:
- New CO switching installations that affect the directly connected OSPs;
- Network router, SR or FE installations, upgrades, or rehomes;
- NPA additions;
- Any other event that affects 9-1-1.

7.4.1 OSP Wireline Including Reseller

SIs have extensive experience in this area and test plan development is time-consuming but routine to them. There are topics unique to wireline that must be considered, including test access numbers to be used from each testing OSP, enterprise, and location. As mentioned in Chapters 5 and 6, the LERG carries test numbers, and almost none of them are accurate or work for testing NG9-1-1. ASR forms are important from each OSP describing the actual connectivity desired and in place before testing can begin. Each originating party must supply full test access information to the 9-1-1 SSP team in advance of any testing.

The ILECs have legacy SRs and end offices serving the public safety agency. Test cases will be executed for each SR and for each end office in the footprint. Configuration information for testing will be recorded prior to the outset of the testing in the tables similar to the ones below. Refer to Chapter 5 and 6.

7.4.1.1 ILECs

The ILECs operating in the public safety agency footprint are identified. Their representatives must confirm that no MF signaling is used. Verify any use of carrier ISDN PRI in the agency footprint. For these ILECs, the POI will be located along the ESInet at the locations indicated in the table shown in the test plan (Table 7.3).

The ILECs may use their own facilities to reach to the POI or use an underlying transport provider. This information will be available to the test team once the ASRs are received. At each POI, a cross-connection must be in place from the OSP colocation point in the building to the ESInet interface using *name brand* routers or equivalent, capable of directing traffic to either data center by design or in an outage or other blocking situation, sending traffic to the alternate data center via the fiber network.

See Table 7.4. The end customer address records and NPA NXXs will be provided by the ILEC into the NG-911 FTP server, in NENA format.

Table 7.3 Points of Interconnection Along the ESInet Table Number 1

			Traffic Type		
Location	*Building (as Labeled on Diagrams)*	*Entity Using Colocations at This Address*	*Inbound Access*	*Outbound Split*	*PS/ ALI*
Full Address of each Legacy SR Location	Name Data Center 1 CLLI	All Access Carriers and PS/ALI	Yes	Yes	Yes
	Name Data Center 2 CLLI	All Access Carriers and PS/ALI	Yes	Yes	Yes
	POIs other than Data Center Name Address and CLLI	ILECs that have interfaces	Yes/No	Yes/No	Yes/ No

Table 7.4 Legacy Selective Routers (Signaling is SS7)

Carrier Name	*Selective Routers*	*Connectivity Locations in Data Centers, on ESInet)*	*Transport Provider*	*Database*
	CLLI	Direct facility to the data center POIs	PSAP SR	MSAG/ LVF
	CLLI	Via these POIs on the ESInet to reach the Data center POI:	ASR	MSAG/ LVF
		Wire center CLLI Wire center CLLI		
		Address of each POI or Data Center Subcontractor Cross-connections to ESInet		

For the end offices that connect with multiple redundant trunks, calls from the ILEC network will be routed using a lower-alpha-first routing plan, such that all traffic will initially be sent to data center 1, with overflow and busy attempts going to data center 2. Routing rules will be discussed with each OSP. The ILECs in the footprint are advised that they will be routing to data center 1, alternate route via data center 2. Future interfaces are planned to the next legacy SR when the legacy SR carrier is prepared to rehome designated split exchanges off the primary legacy SR. Advanced planning is needed to make these changes.

Consider by way of example a wireline CO. If calls are made from a CO location, the telephone numbers and addresses from which calls will originate must be added to the NG9-1-1 database. Often those locations—because they are typical telco locations—do not show up in databases at all because they were not installed by a SO process. Originating test numbers are a specific spreadsheet assignment that cannot be overlooked. If it is necessary for local public safety or civic locations to be used for initiating tests, those names, numbers, and people must be contacted in advance and agree to participate. Often testers are solicited from local fire departments, police stations or ambulance agencies, city halls, and more friendly groups. Testing must come from unique classes of service to prove that 9-1-1 calls are handled correctly. TTY customers or lines need to be tested for compliance. In Split exchanges, calls must originate from locations represented by each area that falls into Split exchange boundaries. If the public safety agency is responsible for calls from waterways and other states, countries, or unique landmarks, plan for testing from those locations. The type of testing samples required is possibly dependent on the oversight jurisdiction if any exists. Negotiation may be required to be reasonable. There are no generally automated programs that can run these types of tests across the board for all customers in an area. In wireline cases, much of the world uses a combination of host and remote switch arrangements. The testers must know that the host remote relationships and ensure test calls are made from each host and remote and that they, of course, end up in the right PSAP. Since this is a new NG9-1-1 database (LVF/LIS), using GIS routing, locations cannot simply be rural routes. It is essential that full addresses show up on a PSAP Map. Because there are so many types of host and remote vendors, every single type and location needs to be tested. We learned that what passes through a legacy SR may not go directly into a NG9-1-1 network without failure. There are so many end-to end-interworking protocols, it is essential to be thorough. Then there is the issue of diverse access testing to each data center for first route and alternate route completion and failover. Busy conditions must be simulated, and the alternate diverse routes or paths must be tested The design and expected outcome must be on the list and verified. In the case of landlines, all are not created equal. ILECs need to follow most of the above listed testing suggestions and rules.

7.4.1.2 CLECs

The facilities-based CLECs serving the footprint often use SS7 as the signaling protocol when connecting to the physical point of interface (POI). The CLEC may use its own facilities to reach the POI to get into the NCGS or use an underlying aggregator. The database lookup functionality may be provided by the CLEC or by a third-party database provider. The locations of the POIs will be available at the completion of the pending ASRs. The information that is known about facilities and database is available from the carrier listings and the ASRs.

The CLECs serving the public safety area usually have at least one switch. Test cases will be executed for each switch serving the agency's footprint. Configuration information will be recorded at the outset of the testing in the Table 7.5. Testing for CLECs that serve the area through a third-party is accomplished by testing with the third-party. Testing of network connectivity and routing between these CLECs and their third-party supplier is outside the scope of this section.

Calls from the CLEC network are routed using a class-of-service mechanism[3]; only appropriate traffic for the ESInet's PSAPs will be sent to the data center equipment. Calls that belong to agencies outside of the NG9-1-1 ESInet area should not appear at either data center. Carriers that use the scheme of alphabetical A–L carrier names will route first to data center 1, alternate to data center 2, and the reverse occurs for M–Z carrier names.

CLECs often can perform remote testing, which will speed the process.

3. Class of service in this case means that a single NPA-NXX may serve a state or large graphic area, not simply a specific rate exchange. CLECs are required to direct their 9-1-1 callers to the appropriate PSAP using the full XXXX line numbers of the caller. CLECs do much more originating call translations and have more 9-1-1 trunk groups in place for sending calls to the correct PSAP. This work is not required by a similar ILEC or RLEC. CLECs are not allowed to use split-exchange routing.

Table 7.5 Sample CLEC Tracking

CLEC	*Sig Type (SS7, MF, SIP, ISDN PRI)*	*Connectivity Location (In Data Centers, On ESInet)*	*Aggregator (may be none)*	*Database*
Name	SS7	ASR	ASR	ASR
Name	SIP	ASR	ASR	ASR
Name – Indicate (transport)	SS7	ASR	ASR	Name VPC

7.4.1.3 Resellers and PS/ALI

If an OSP is a landline reseller, the entity it purchases service from is responsible to bring the OSP into the testing process and be sure that it conforms and its clients reach 9-1-1 after the cutover. That is also true of any enterprise using PS/ALI. Most large enterprise cutover activity will likely be delayed until after all OSPs are connected. Thus, the legacy SR interfaces for access remain in place until the last customer is tested and cut over. Those SPOF legacy SRs are not going anywhere for a while, and in fact they will have new and updated interfaces, tests and transfers uses.

7.4.2 OSP Wireless

The wireless OSPs serving the public safety agency have at least one MTSO and send traffic by way of one or more MPCs. Test cases will be executed for each MTSO in the footprint. Configuration information will be recorded prior to the testing. ESRK information will be verified before testing begins. See Table 7.6.

The wireless OSP may use SS7 or SIP as the signaling protocol when connecting to the system. The POI may be located along the ESInet, or at data center locations, which will become available at the completion of the ASRs. The OSPs may use their own facilities to reach to the system or use an underlying transport provider. The wireless OSPs in the area support texting to 9-1-1. Calls from the MTSO network will be routed based on the location of the caller. Agency traffic will be sent to the data centers. Calls that belong to agencies outside of the ESInet area should not appear at either data center. Carriers will use the scheme in which alpha A–L carrier names route first to data center 1 and alternate to data center 2 and the reverse for M–Z carrier names. Note that carriers using some aggregators will, for example, route to the data center 1 first; consider the aggregator's alpha name.

Testing wireless involves verifying or exchanging ESRK and tower information and any tower face or other identifiers with the ASR. Persons may be using the

Table 7.6 Wireless Access Testing

Wireless Carrier	*MPC*	*Sig Type (SS7, SIP)*	*Connectivity Location (in Data Centers, on ESInet)*	*Transport Provider*	*Texting (y/n)*
Name	ASR	SS7	ASR	ASR	Y
Name	ASR	SIP*	ASR	Aggregator	Y

* Wireless signaling is SIP-to-aggregator in some cases, which reoriginates as SS7 into the NCGS.

wireless OSP's phones and traveling around the area to be sure that calls land in the right PSAP and can be transferred as callers move in their vehicle or on foot. Map locations are critical.

7.4.2.1 Resellers: Wireless

It is important to realize that many wireless providers operate in a territory under different names than the primary OSPs. Think of them as resellers for wireless. Make sure that each reseller are contacted by the primary and tested through the OSP similar to the reseller strategy mentioned in Section 7.4.1. In this case, it is prudent to test from an OSP that is not officially in the footprint but that has customers driving though to see what comes up on the maps and to verify the locations that arrive.

7.4.3 OSP VoIP

VoIP OSPs that will be sending NG9-1-1 traffic into the footprint will more often than not use an aggregator. Fixed-VoIP OSPs remain a possible exception, and that may include cable TV providers or ILECs that offer bundled service as referenced in Chapter 5. The POI may be located along the ESInet, or at the data center locations. The VoIP OSPs will use the scheme of alpha A–L carrier names routing first to data center 1, alternating to data center 2 and the reverse for M–Z carrier names. Test plans for VoIP providers look exactly like the plans for the wireless providers. The major VoIP aggregator contacts will be made and the testing will proceed, ensuring that the VPC-provided location information is accurate when the calls arrive. Historically this has been completed quickly since the third-party groups tend to use the same facilities provider in the United States, and they split up representation of almost all the VoIP OSP 9-1-1 service. The dominant aggregators still use SS7 signaling for access, so integration testing and interoperability will be key. SIP direct testing to and through the POI to the SBC and then NGCS to the correct PSAP testing may be possible, but from experience we have not seen this step complete. There are dozens to hundreds of VoIP OSPs even in a regional footprint. This is a matter of setting up the testing appointments and personnel and moving through the steps.

The OSPs may use one or more VPCs or utilize fixed-VoIP locations. Test cases will be executed for each VPC. Configuration information will be recorded at the outset of the testing in Table 7.7. This information as well as the physical locations that Level 3 or others like them, connect becomes available at the completion of the pending ASRs.

Table 7.7 VoIP Access Testing

VoIP Carrier	*VPC*	*Sig Type (SS7, SIP)*	*Connectivity Location In Data Centers, On ESInet)*	*Transport Provider*
Name	ASR	SS7	ASR	ASR
Name	ASR	SIP *	ASR	Aggregator

* VoIP signaling is SIP-to-aggregator, which reoriginates as SS7 into the NCGS; at a future date, SIP will require retest.

7.4.4 Enterprise Direct

PS/ALI deployment has two types of traffic, directly connected PS/ALI customers routing through the legacy SR, and PS/ALI customers routing through the local switch. This example enterprise uses a DMS100 switch with SDN PRI as the signaling protocol to reach the NGCS. For all other individual PS/ALI customers routing through the legacy SR, the signaling information is out of the scope of this test plan as it occurs behind the legacy SR. The signaling from the SR into the NGCS is SS7. The calls will route through the legacy SR POIs to the two data center POIs until individual contracts can be arranged with each and every required PS/ALI enterprise customer. This involves legal steps, and the ILECs, RLECs, and major CLECs with these types of customers have not been forthcoming to work on NG9-1-1 as of this book's publication. Chapter 11 further discusses these issues. Since the legacy SR is going to be connected to the diverse data centers, and trunking will most likely be changed to SS7—as we have not seen a legacy SR capable of SIP—it is advisable to make at least a few test calls to make sure the database for location is still operating properly for dorms, hospital rooms, remote campus locations, and major government entities and that everything is flowing through to the PSAP properly. Local testing rules and guidelines may apply. While this should operate the same as any other traffic remaining on the legacy SR, these are very large customers, and any problems can become significant. See Table 7.8.

Each participant in the testing will provide configuration information to the 9-1-1 SSP, which should be maintained throughout the testing period and saved for future use.

7.4.5 OSP Configuration

ASRs may carry this level of detail for each OSP and PS/ALI customer. It is essential not only to know this information but also to have these facilities in place and the ESInet and NGCS able to accept the terminations well ahead of scheduling testing. Allow for up to 90 days from order to installation in the project timeline. Connection agreement contracts must be in place before any OSP or enterprise PS/ALI customer will issue an order for facilities, especially to a new data center(s). The 9-1-1 SSP can request diverse access, but not usually mandate it. Testing for failure is more important than ever if diversity is not provided. This wait time is sensitive

Table 7.8 Enterprise PS/ALI Access

PS/ALI Customer	*Switch*	*Sig Type (SS7, SIP,ISDN)*	*Connectivity Location (in Data Centers, on ESInet)*
Name	Legacy SR	SS7	Via these POIs on the ESInet to reach the data center POI:
Name	Legacy SR	SS7	
Name	Legacy SR	SS7	Wire center name—Wire center name
			Addresses of diverse POIs or data centers
			When they switch to the direct interface many are IP PBXs and may wish to use SIP.
Major college	Sample campus switch	ISDN PRI	Customer is already connected On ESInet: at data center

if SS7 is involved and needs added testing and connectivity for A links, facilities, and trunking.

Each participant in the testing will provide additional configuration information to the 9-1-1 SSP, which will be maintained throughout the testing period, and saved for future use.

7.5 End-to-End

At this point, the parts of the new NG9-1-1 network, the transport and the PSAPs, and access from all sources has been tested. Now it is time to address the actual end-to-end testing plan. This is where all the individual tests come together to prepare for cutover. For each place where there were problems during the prior testing process, the project manager would have sought resolution and determined how to resolve or work around the issue. As in earlier tests, the actual PSAP personnel were not getting the test calls. They were not making transfers or ticketing the troubles. This was all part of the test team responsibilities.

The end-to-end test plan is the assembly of all the test plans into the same spreadsheet. The project team leaders can review the plans, and based on lessons learned determine how long it takes to get the PSAP personnel trained and the parties assembled for the cutover. Maintenance windows were not an issue before because no live traffic was impacted. Now the plan must be organized and sequenced to fit the maintenance windows. Real MOPs must be written, and the draft plans shared with all personnel who have any part of the cutover.

7.5.1 Test Cases

The functional test checklist is included in this section. Each item will be tested based on the availability of test circuits and interfaces.

7.5.2 Requirements Codes

The following codes indicate pass, fail, or not applicable for each feature. The designated code for each requested feature is listed in the "test code" column.

TP	Test passed
TF	Test failed
RTP	Retest passed
RTF	Retest failed
NA	This feature is not applicable in customer systems

When the TF or RTF codes shown above are used, the tester must also insert an explanation of the planned correction action in the comments column.

The following test cases support various network and OSP configurations. The test cases that are relevant to a particular type of access carrier facility can be tested as those particular facilities are activated. If feasible, the test cases that represent a facility type for which there are multiple access carriers will be repeated for each carrier.

For information on the physical location of connection point. Upon completion of pending ASRs, additional information will be available. Complete Table 7.9 for each OSP. All access transport tests will include a colocation carrier where appropriate. See Table 7.10.

7.5.3 Signaling Test

This section shows signaling related test cases. Depending on the type of signaling used by the OSP some of these test cases may be out of scope. Test cases that are not applicable will have "NA" in the results column. Note that where test cases

Table 7.9 Access Termination

Connection Type	*Comments*
Transport type	
Facility type (fiber or copper)	
Termination capacity (DS1, DS3)	
Termination location	
Database provider or FTP server	
Legacy SR interface to data center (split exchange carriers only)	
Signaling type	
Numbers of trunks per end office, host, or MTSO	

Table 7.10 Transport Test Cases

Test Case #	*Test Description*	*Expected Test Outcome Notes*	*Results/ Date*
NG-TRN-001	Demonstrate the ability to connect from the carrier network to the POI.	Carrier can connect to the ESInet or to a POI in the data center; bandwidth matches the plan.	
NG-TRN-002	Demonstrate the ability to connect from the carrier network to the data centers.	Carrier can connect to both data centers; bandwidth matches the plan.	
NG-TRN-003	Demonstrate the ability to connect from the carrier network to NG core equipment at the data centers.	Sessions can be established to the LNG LSRG, BCF, or ESRP, or LIS	
NG-TRN-004	Carriers using access point not on the ESInet—Demonstrate the ability to connect to and through the colocation point on the network through to the data center 1.	Carrier can connect and with stated bandwidth to the colocation point.	
NG-TRN-005	Carriers using access point not on the ESInet—Demonstrate the ability to connect to and through the colocation point on the network through to the data center 2.	Carrier can connect and with stated bandwidth to the colocation point.	
NG-TRN-00-006	Demonstrate calls going to intermediate ESInet POIs through Colocation interfaces through the can reach both Data Center POIs		

NG-SIG-006 through 009 show "system STP pair," that refers to the STP pair of the SS7 service provider that is a partner to the 9-1-1 SSP.

Each carrier's SS7 PCs for each switch, wireless or wireline, location, and STP pair and locations will be provided to the 9-1-1 SSP and SS7 provider subcontractor prior to network testing and used during the OSP testing. Those records will be maintained by the 9-1-1 SSP.

Use Table 7.11 for SS7 test cases. Use Table 7.12 for routing and failover test cases. Use Table 7.13 for split exchange testing, to be provided by each carrier prior to testing. (Each carrier also will verify NPA NXX or alternate numbers and add to the list before tests begin.)

Use Table 7.14 for agency directory numbers. For traffic split by the NG 911 system, Table 7.15 serves as an example.

For traffic split by the legacy SR, see the Table 7.16.

7.5.4 Texting Test Cases

Testing text-to-911 involves the use of cell phones with text capability using the wireless carriers' services. More test cases will be added as contracts and other agreements are completed between the 9-1-1 SSP and various carriers and partners. TCCs may have additional access tests that can be added below as required and duplicated as each county is added. See Tables 7.17–7.20.

7.5.5 Testing SLAs

NOC Information will be exchanged for use after Cutover. The SLAs are specified in the Legal Agreements between NG-911 SSP and each Carrier.

This section identifies specific testable aspects of the SLA or carrier connectivity agreement between the carrier and the 9-1-1 SSP. The reference for this section is the NENA Standard Document for Design 08 506. The Glossary of Terms and Acronyms has terms defined.

The testing will validate the measurements, as set forth by NENA and 9-1-1SSP validate that the measurements, as set forth by NENA and 9-1-1SSP are within are within design limits between the access carriers through the ESInet to the PSAPs.

Test Status:	*Type of Testing and Results*	*Pass or Fail*	*Notes*	*Date*
To Be Done	Packet Loss			
To Be Done	Jitter			
To Be Done	Latency			
To Be Done	Other			

Obstacles don't have to stop you. If you run into a wall, don't turn around and give up. Figure out how to climb it, go through it, or work around it.

—Michael Jordan

Table 7.11 Signaling Test Cases

Test Case #	*Test Description*	*Expected Test Outcome Notes*	*Results/ Date*
NG-SIG-001	Demonstrate the ability to support multiple types of inbound and outbound analog or digital CO or end office provisioning, such as SIP, SS7, ISDN PRI, and MF.	Based on collected results of individual tests with carriers using different signaling types. The individual test results appear in this table in the following test cases a–d. See below for cases a-d	
NG-SIG-001a	Demonstrate the ability to support SIP calls	SIP call received by the SBC forwarded and received at the ESRP	
NG-SIG-001b	Demonstrate the ability to support SS7 calls	Link set up across SS7 facilities and SS7 call received at LNG, converted to SIP, sent to the SBC, forwarded and received at the ESRP	
NG-SIG-001c	Demonstrate the ability to support MF calls	MF call received at LNG, converted to SIP, sent to the SBC, forwarded and received at the ESRP	No MF for agency
NG-SIG-001d	Demonstrate the ability to support ISDN PRI calls	ISDN PRI call received at LNG, converted to SIP, sent to the SBC, forwarded and received at the ESRP	
NG-SIG-002	Demonstrate the ability to terminate call from the origination using non-emergency treatment	Call drops cleanly	
NG-SIG-003	Demonstrate the ability to terminate call from the data center using non-emergency treatment	Call drops cleanly	
NG-SIG-004	Demonstrate the ability to terminate call from the origination using emergency treatment	Call stays up on the data center end	
NG-SIG-005	Demonstrate the ability to terminate call from the data center using emergency treatment	Call drops cleanly	
NG-SIG-006	If SS7, demonstrate that the carrier STP pairs send signaling messages to the system STP pair	See SS7 configuration information such as PCs	
NG-SIG-007	If SS7, demonstrate that the system STP pair sends messages to the appropriate carrier STP pair	See SS7 configuration information such as PCs	
NG-SIG-008	If SS7, demonstrate that the system STP pair sends the correct network management signals to the appropriate carrier STP pair when the first route is unavailable	See SS7 configuration information such as PCs	
NG-SIG-009	If SS7, demonstrate that the system STP pair sends the correct network management signals to the appropriate carrier STP pair when the first route is restored	See SS7 configuration information such as PCs	
NG-SIG-010	Reserved for future versions		
NG-SIG-011	Generate traffic to the legacy SR and verify that the LSRG converts SIP to SS7	See SS7 configuration information such as PCs	

Table 7.12 Routing and Failover Test Cases

Test Case #	*Test Description*	*Expected Test Outcome Notes*	*Results/ Date*
NG-RTE-001	Demonstrate the ability to establish a call from the carrier to the first route.	Lower alpha or class of service used properly	
NG-RTE-002a	Demonstrate the ability to establish a call from the carrier to the alternate route (make first route busy).	Failover performed properly	
NG-RTE-002b	Demonstrate the ability to establish a call from the carrier to the alternate route (take NG core service element offline).	Failover performed properly	
NG-RTE-002c	Demonstrate the ability to establish a call from the carrier to the alternate route (take first data center offline).	Failover performed properly	
NG-RTE-003	Demonstrate that the system returns the appropriate response when the first route is unavailable.	Overflow performed properly	
NG-RTE-004	Demonstrate that the system returns the appropriate response when the second route is also unavailable.	Network busy returned properly	
NG-RTE-005	Demonstrate that the system returns the appropriate response when the disaster recovery route is also unavailable - This test cannot be performed until after agency n is online for agency 1; note that as of the agency 1 filing all tests between the agency 1 and agency 2 PSAPs are void until an IGA can be reached.	Network busy returned properly	
NG-RTE-006 -NG-RTE-010	Reserved for future versions.		

Table 7.13 Split Exchange Test Cases

Test Case #	*Test Description*	*Expected Test Outcome Notes*	*Results/ Date*
NG-SPL-001	Demonstrate the ability to route calls that belong to agency to the NG core services in the system data centers.	See Table 7.14 for a list of numbers that belong to agency.	
NG-SPL-002	Demonstrate the ability to route calls that *do not* belong to agency to the appropriate county using NG9-1-1 technology (LSRG).	ANI/ALI will be intact for these calls.	
NG-SPL-003	Demonstrate the ability for the 911 SSP that is responsible for splitting the calls to route to 9-1-1 SSP those calls that *do* belong to ESInet area PSAPs. These calls are routed using E911 legacy SR technology.	Refer to the split exchange list for the calls that are the responsibility of other carriers to split. Calls terminate in the correct PSAPs (connection agreements).	
NG-SPL-004–NG-SPL-010	Reserved for future use.		

Table 7.14 Agency Directory Numbers

CLLI Code	NPA (NXX)-NXX	Host	Rate Center	Legacy SR(s)	Test Numbers	Carrier	Type	LATA Name	LATA Num
	NXX	CLLI	Name	CLLI, CLLI	3 Needed	Name	ILEC	LERG Name	LERG Number 3 digits
	NXX		Name	CLLI	2 Needed	Name	ILEC	LERG Name	LERG Number 3 digits
	NXX	CLLI	Name	CLLI	2 Needed	Name	ILEC	LERG Name	LERG Number 3 digits
	NXX		Name	CLLI, CLLI	3 Needed	Name	ILEC	LERG Name	LERG Number 3 digits

Table 7.15 Legacy Selective Router Table

Legacy SR	*Subcontractor STP*	*NG 911 LSGR*	*PSAP*
CLLI	CLLI	Data center 1	Sheriff, PD, neighboring county sheriff
CLLI	CLLI	Data center 1	Sheriff, PD, neighboring county sheriff
CLLI	CLLI	Data center 2	Sheriff, PD, neighboring county sheriff
CLLI	CLLI	Data center 2	Sheriff, PD, neighboring county sheriff

Table 7.16 SS7 Provider Table

Legacy SR	*STP*	*NG 911 LSGR*	*PSAP*
CLLI	CLLI	Data center1	Sheriff, PD, neighboring county sheriff
CLLI	CLLI	Data center1	Sheriff, PD, neighboring county sheriff
CLLI	CLLI	Data center 2	Sheriff, PD, neighboring county sheriff
CLLI	CLLI	Data center 2	Sheriff, PD, neighboring county sheriff

Table 7.17 Texting Test Cases

Test Case #	*Test Description*	*Expected Test Outcome Notes*	*Results/ Date*
NG-TXT-001	Demonstrate the ability to deliver a text message.	Text is received by NG core.	
NG-TXT-002	Demonstrate the ability to reply to a text message.	A reply can be returned to the texter.	
NG-TXT-003	Demonstrate the ability to deliver a text message to the alternate data center when the primary is unavailable.		
NG-TXT-004	Demonstrate the ability to reply to a text message from the alternate data center when the primary is unavailable.		
NG-TXT-005	Demonstrate the text messages are recorded properly.	Texts are recorded, retrieved, and with proper timestamps.	

Table 7.18 ANI/ALI Test Cases

Test Case #	*Test Description*	*Expected Test Outcome Notes*	*Results/ Date*
NG-ALI-001	Demonstrate the ability of the switch or MTSO to deliver ANI and automatic location from the NG9-1-1 database in the datacenters or via the VPC/MPC databases in the network for mobile and VoIP plus some CLECs that choose that solution for location. Identification (ALI) data display to the ECRF.	Receive ANI/ALI.	
NG-ALI-002	Validate that NG Core Services equipment can retrieve ALI equivalent information from database system.	Can retrieve ALI information.	
NG-ALi-003	Validate that NG core services equipment can provide for ALI retrieval from Caller ID or manually entered telephone number.	Can retrieve ALI information.	
NG-ALI-004	Verify connectivity to LIS.		

Table 7.19 PS/ALI Test Cases

Test Case #	*Test Description*	*Expected Test Outcome Notes*	*Results/ Date*
NG-PSA-001	Demonstrate the ability to establish a call from the PS/ALI customer to the first route into the NGCS.	Lower alpha or class of service used properly.	
NG- PSA -002a	Demonstrate the ability to establish a call from the PS/ALI customer to the alternate route (make first route busy).	Failover performed properly.	
NG- PSA -002b	Demonstrate the ability to establish a call from the PS/ALI customer to the alternate route (take NG core service element offline).	Failover performed properly.	
NG- PSA -002c	Demonstrate the ability to establish a call from the PS/ALI customer to the alternate route (take first data center offline).	Failover performed properly.	
NG- PSA -003	Demonstrate that the system returns the appropriate response when the first route is unavailable.	Overflow performed properly.	
NG- PSA -004	Demonstrate that the system returns the appropriate response when the second route is also unavailable.	Network busy returned properly.	
NG- PSA -005	Demonstrate that the floor and building number of the caller match the location received.		
NG- PSA -006	Demonstrate that the caller's location appears correctly on the map.		
NG- PSA -007	Demonstrate that all PS/ALI customers can complete 911 calls.		
NG- PSA -008	Demonstrate that all PS/ALI customers' 911 calls route to the appropriate PSAP.		

Table 7.20 Site Acceptance Sample

All tests were executed and passed, including those in Appendix A		
All tests were executed, with the following deviations:		
Test Case Number	*Deviation and Acceptability*	*Future Resolution Plan*
We hereby certify that the present document is complete (no missing pages) and that all test procedures have been executed and passed as per associated expected results or that all tests have been executed and that some deviations were observed.		
NG9-1-1 SSP Test Representative/Witness		
Name:		
Title:		
Signature:		
Date:		
Carrier Test Representative/Witness (includes acceptance of third-party results where applicable)		
Name:		
Title:		
Signature:		
Date:		

Selected Bibliography

NENA-INF-008.2-2014 (originally 77-501), NG9-1-1 Transition Plan Considerations.

CHAPTER 8

Cutover

8.1 Overview

The success of the cutover to NG9-1-1 is measured by the amount of disruption experienced, which ultimately determines the success of and satisfaction with the entire project. Converting from an E9-1-1 network to a NG9-1-1 network is not only a circuit, facility, and equipment change, but also a software and database cutover. This requires extensive planning, coordination, communication, record-keeping, and troubleshooting. Often NG9-1-1 cutover projects involve changing the 9-1-1 SSP, which can add a level of complexity to the project since it implies that the incumbent 9-1-1 SSP lost the business. The cutover strategy is the same even if there is no change in the 9-1-1 SSP.

A primary consideration is to minimize disruption of operations. Operations will be impacted during the cutover. Steps the project team must take to ensure that no 9-1-1 calls are dropped, misrouted, or sent to fast busy or intercept announcement are listed as follows:

1. Determine basic strategy for cutover;
2. Develop a cutover plan;
3. Schedule the cutover;
4. Pressure-test the plan (i.e., review the plan frequently with a critical eye to security);
5. Communicate the plan;
6. Execute following the plan;
7. Fix what is broken or unexpected;
8. Review the cutover—good and bad—to learn what to do next time;
9. Perform post-cutover legacy system removal.

8.2 Determine Basic Strategy for Cutover

There are a few tactics that should be deployed across all strategic options; these fall under project management and best practices.

The project manager identifies the necessary personnel and documents their contact information to be shared with the team. Each team member knows their role(s) to pressure-test the cutover plan in formal tabletop exercises. Tabletop exercises are simulations that help to identify team members who must be present at the cutover or on the cutover conference calls and those who merely need to be available on a standby status. Tabletops are used to walk through every step of the cutover plan.

It is critical to have at least one published voice/video bridge open during the cutover and a second voice/video bridge available if problems arise. The technical team needs one voice/video bridge to implement the well-documented plan; administration, management, and vendors use one voice/video bridge for command and control decisions.

In terms of best practices, the authors recommend reviewing the times the PSAP has the least traffic for scheduling the start of the cutover, and, depending on the specific strategy, determine cutover windows in more protracted cutovers. This will most likely result in a morning-to-afternoon weekday cutover window. In the telecom industry it was and is still common to cutover PBXs, switches, and major facilities and trunking in the middle of the night on weekends. This is done because of typically low traffic volumes for commercial businesses. Public safety and 9-1-1 operations find that their traffic volumes are busiest on a different schedule than commercial businesses. PSAP busy time often peaks in the evening-to-overnight period, especially on weekends and holidays. The bonus to cutting on a weekday is that the cutover team has the most experienced people working at the OSPs, including management and administrative personnel who are readily available during normal business hours.

Target periods for cutover may be in low-traffic times, but there is never a "no-traffic" time. Also, failover functionality means that the alternate PSAPs have to be ready to take the calls, and that means staffing agreements are in place. The new functionality includes transfers. That means that PSAPs on-net and off-net and agencies such as sheriff offices and state police must be ready to accept the transferred calls. OSPs do not have unlimited personnel, and not all test calls for the cutover can be originated remotely. Often a human must drive to locations to assure that the caller location shows up on the PSAP map and is identifiable for dispatch.

We cannot leave anyone without 9-1-1 service, and wireline OSPs cutovers are where leaving someone without 9-1-1 is most vulnerable. During cutover testing, the TTY equipment must be tested from each exchange or CO as well. The cutover spread sheets, which stem from preceding work done in Chapter 7, need all variables listed.

There are two main categories of cutovers, with NG9-1-1 adding a third; they are described as follows.

- *Flash cutover:* With a flash cutover, a literal or figurative switch is thrown, and *all* the traffic moves from the legacy E9-1-1 network to the NG9-1-1 network. A longer variation of the flash cutover means that instead of moving all the traffic to NG9-1-1 at once , the team cuts one trunk group over dedicated facilities at a time, testing each facility and trunk group, then cutting the next one in sequence. This process takes longer but is far more reliable in getting all the traffic moved to the NG9-1-1 system. "Why is this called a flash cutover?" you may ask. It is cutting the traffic on a certain date and time based on a schedule that allows for call-through-testing one group at a time, and if any problems exist, the alternate PSAP(s) can handle the live 9-1-1 calls until the testing and resolution of problems can occur. Of course, every group will have been pretested prior to cutover, but even in a short interval something can change in the network, such as a facility roll or switch generic upgrade that alters success. The team can never be too careful with 9-1-1 traffic.

- *Parallel cutover:* A parallel cutover is where the legacy E9-1-1 system remains fully functioning, the new NG9-1-1 system is installed in parallel, and both remain operational until there is assurance of a completely successful cutover. This is far more reliable but also more expensive, as the legacy system needs to stay in place for some period of time, generally from a week to a month. While this is possible for the core network, the originating and terminating ends seldom can be totally duplicated. In many cases new facilities and trunk groups need to be installed to the new ESInets. However, the actual trunking in the OSP origination locations will be reused to terminate to the NG9-1-1 network at the time of cutover. In the PSAP, it is seldom possible to have a complete set of duplicated PSAP workstations to operate in parallel with workforces straddling both systems for an extended period of time. A few workstations can be set up to take the new calls while the old calls are being answered by the staff operating the E9-1-1 positions. Once the call-through testing is complete, the onsite cutover personnel can move additional workstations in place until they are all transitioned. The 9-1-1 call volumes start out small as one OSP at a time swings their traffic.
- *NG9-1-1 hybrid cutover:* Cutting to a NG9-1-1 network requires both a flash cut and a parallel cut concurrently. Because of NG9-1-1 having multiple OSPs directly connected, it takes time to get all OSPs' traffic cutover to the new NG9-1-1 system. The OSPs are individually flash-cut, but the ones queued to cutover are in a parallel cutover mode.

8.2.1 Strategies

Some alternative cutover strategies:

- SR(s) cutover: A diverse route to each data center;
- Single public safety agency/multiple PSAPs direct connect carrier or OSP cutover:
  - Single PSAP at a time;
  - Single OSP or carrier at a time.
  - Regional NG9-1-1 project direct connect carrier cutover:
  - Single PSAP at a time;
  - Single OSP or carrier at a time.

8.2.2 SR Traffic Cutover

Systems where IP-based 9-1-1 equipment is installed in PSAPs and the traffic is still coming from a SR would resemble E9-1-1 conversions and involve cutting the CAMA (or equivalent) trunks to POIs at the ESInet and the facilities extended into the BCFs. (Refer to Chapter 3.) While this is the simplest and most straightforward type of cutover, careful coordination must occur using the techniques outlined in this chapter. Often, if not always, in this situation, the ALI database is controlled by the owner of the SR, even though it may be housed by a third party—typically a telco SR—and the SO updates comes from the same source. PSAP space considerations must be taken into account as the old equipment needs to remain to handle live traffic from the uncut trunks at any point in the cutover, and at least one workstation should remain live on the legacy network to catch any traffic that slips through after the cutover is complete.

New trunk groups between the legacy SR and the NG9-1-1 network must be installed and tested ahead of any planned cutover activity. These new groups should not be using CAMA trunks, and while SIP trunks over Ethernet are preferable, followed by ISDN-PRI, we know from experience that the trunks are SS7 most of the time because that is what is technically available. Those trunk groups are important even past cutover due to the likelihood of transfers needing to use those paths to adjacent agencies and due to the LSGR interface requirements for split exchanges and enterprises that are not likely be cut over at the same time as OSPs. This is a lot to absorb without reading prior chapters of the book.

Another reason that SR legacy trunking might be needed is because the OSPs and enterprises refuse to cut over direct trunking when the area converts to NG9-1-1 or because there is a phased strategy to cut over the NG9-1-1 network and PSAPs leaving the OSP side tethered to the SPOF SR just to have one end of the cutover managed and have fewer moving parts at once. This phased strategy leaves at least two 9-1-1 SSPs operating in the same area. This interim solution actually introduces added time to reach the PSAP and does nothing to help the bottlenecks the legacy SRs pose to the free flow of 9-1-1 calls. This process costs more for the agencies and causes complex contract negotiations among parties. That said, it has been used frequently because challenging negotiations for direct connection to new data centers pose issues with the originating OSPs.

8.2.3 Single Public Safety Agency—Multiple PSAPs Direct Connect Carrier Cutover

A true N9-1-1 network eliminates the SPOF inherent in SR and E9-1-1; redundant hardware and network infrastructure is an improvement, but by directly connecting with the OSPs, the reliability factor is improved end to end, and the ongoing operating costs are reduced for nearly all parties, except that the existing 9-1-1 SSP loses revenue. In today's competitive world of technologies and multiple OSP choices, there could be as few as 10 or 12 OSPs to deal with or as many as 200 depending on the size of the public safety agency. Obviously, population attracts the OSPs to a market, so large cities have the maximum number of OSPs, while rural public safety agencies and their markets have fewer. Ironically, in rural environments there may be more wireline-based local-exchange OSPs to interface. These wireline OSPs are often what were called independent telephone companies or sometimes mom-and-pop phone companies.

This hybrid solution is our preferred method of cutover. It involves more scheduling and timing with OSPs ahead of the cutovers, but limiting the cutover to a handful of PSAPs in one agency's area can be the most cost-effective and avoids going back to phased approaches. One way to accomplish this is choosing to cut over classes of OSPs by day. Since approximately 80% of the traffic comes from wireless and VoIP OSPs, it is helpful to schedule the wireless carriers first. There are fewer of them, and this allows for a rigorous test of the ESInet, data centers, NG9-1-1 NGCS, and the PSAPs. Depending on the PSAPs, they can estimate staffing to be ready and the number of new positions needed for the call volumes, because they know their volumes by not only time of day and day of week, but also by type of OSP. Most of the calls requiring transfer come from wireless callers who are typically the ones on the move. The following cutover day(s), the VoIP and CLECs using VoIP-like access can be scheduled. This results in those hybrid callers who think that their phones are still landline but have bought bundled service from a cable TV provider to be cut over. Managing the schedule for those providers should be reasonably straightforward. At that point it is likely that only approximately 10% of the most complex callers, wireline customers, are left to migrate. They could be left on the old SR for a week, but that means that the PSAPs would need to manage two types of positions for an added week. So, if a third day in a row is available, those wireline customers should be migrated. This could still involve as many as three or more major carriers or OSPs. Seldom do these major OSPs have staffed COs where test calls need to originate. Specific scheduling is required. At this point split exchanges become time-consuming to set up cutover test calls and ensure success.

8.2.4 Small Footprint—Single Public Safety Agencies, One PSAP at a Time

This strategy, which would cut all the OSPs for each PSAP and not cut the next PSAP until complete, would tend to work best if certain call types went to different PSAPs. For example, in some public safety agencies, all wireless and VoIP calls go to the sheriff's office as well as wireline calls and fixed VoIP. If the other PSAPs receive wireline calls and fixed VoIP, then the PSAPs can be cut one at a time since there are discrete OSPs in each PSAP. The strategy suggested would be to cut the wireless OSPs first, starting with the wireless OSP perceived to be the largest and cut in order from largest to smallest. The third-party providers with OSPs have not been willing to disclose any market share information, but an educated guess can be made. This will move the most traffic. The wireless OSPs should make test calls from each tower face to ensure that the call routes to the correct PSAP. Depending on the size of the footprint and number of towers, this can take considerable time depending on the resources the wireless OSP devotes to the cutover. The next steps would be to cut the fixed-VoIP OSPs and the VoIP-type OSPs that use third-party services. The wireline calls should be saved until the end. The complexity of split exchanges and the number of COs that need to be tested takes a significant amount of time for the relatively small percentage of the traffic.

8.2.5 Large Footprint Cutover—Multiple Public Safety Agencies (Regional), One Carrier at a Time

This strategy would be to cut one OSP at a time across all the PSAPs and not move to the next OSP, and so on. This strategy would essentially flash-cut each OSP across the ESInet area, and all traffic for each OSP would enter the NG9-1-1 system. It would also be a parallel cut in that this strategy may take several days, and 9-1-1 calls would come in from both the new NG9-1-1 network and the legacy E9-1-1 network to each PSAP. The advantage to this strategy is that there are not multiple cutovers with the same OSPs to each PSAP. The same suggestion of cut order for the OSPs would be to cut the wireless OSPs first to move the most traffic to the NG9-1-1 system with the smallest number of wireless OSPs in the footprint. Follow the wireless OSP cutover with the fixed VoIP and third-party VoIPs and finish with the wireline OSPs. A regional NG9-1-1 project may eliminate many but not all of the split exchanges as the embedded wireline OSPs send all the traffic to the NG9-1-1 network, and the location-based routing component of NG9-1-1 resolves the split issue. Split exchange handling with adjacent agencies not in the regional project needs to be coordinated with the neighboring agencies and OSPs. Refer to Chapter 5.

This approach to a cutover is easiest for each OSP and harder to coordinate across a region or across multiple agencies. The timeline to deploy this scenario is significantly longer as well, because each agency in a region must have the NG9-1-1 core and i3PSAPs installed and tested and, if necessary, be approved by a regulator. More personnel on the PSAP side would be required for a simultaneous effort, and a troubleshooting team and PSAP workstation team would need to be deployed across a regional footprint, which might stretch resources. A OSP may or may not offer service across all the agencies in a region, and if there are landline OSPs, an even larger team must be deployed. Even wireless OSPs require more people with using test equipment across a large footprint. Everyone connected inside the ESInet and outside would need personnel deployed at the same time to be sure cutover testing of transfers to participating and adjacent agencies works. This type of deployment is massive and might take months to find adequate times to get the testing done in the least busy days and times. There are some regions that are large, yet have only a major PSAP and a disaster-recovery PSAP, such as in major cities with millions of residents. Likely that is the type of location conducive to this type of cutover strategy.

8.3 Develop a Cutover Plan

The cutover order of the PSAPs needs to be researched and planned carefully using fact-based evidence and reasoning in the determination of the cutover schedule. The documentation gathered in Chapters 2 and 4 is useful to determine the cutover strategy and order of the PSAPs, where applicable, and the involved parties, including the neighboring counties and all OSPs. The proposed schedule should be reviewed with the project team to discuss the pros and cons of the cutover strategies, order, and schedule. Whenever the team sets a schedule, be assured that it will be opposed by some party. Depending on who raises the objection to the schedule, assuming it is not such a powerful involved party to influence the project, it should be an argument of evidence and reasoning versus politics and power plays.

8.4 Schedule the Cutover

8.4.1 911 Data Preparation—ALI/GIS

The database preparation takes the longest timeframe of the project to complete and should be started as soon as the project is initiated. The GIS/MSAG and ALI database must be tested and ready prior to the planned cutover. Assuming that the databases are ready and tested and that the daily SOI files have been processing properly for several weeks, the cutover scheduling can begin. The project manager should verify that SO changes are being delivered to each OSP via the predetermined methods agreed upon in contracts. Some of these methods may include, but are not limited to FTP downloads, email, and orders provided through an established website. Check the regulatory rules or customs of the individual states to determine whose responsibility it is to provide and publish the LVF information. Contract agreements should have established the roles' responsibilities, and the processes need to be validated well ahead of final cutover.

8.4.2 Database SOI Updates—Ongoing

As discussed in Chapter 3 there should be concurrent SOI updates from the legacy 9-1-1 OSPs and PS/ALI customers' databases with the identical updates processed into the NG9-1-1 database for about 10 days. After the databases are compared, and the error ratio shows less than 1% of unexplained differences, the NG9-1-1 LIS database is ready for cutover. All exceptions must be manually reviewed and resolved in the timeframes allowable per state statute or SLAs.

Eventually, the LIS database is loaded and tested. The daily SOI files have been processed for several weeks. LIS changes are being delivered to OSPs via the predetermined methods established.

Once each PSAP is cut over live and test calls are made, the 9-1-1 SSP and public safety agencies will ensure that ALI information is displayed on the PSAP workstations and that wireless calls are handled correctly (i.e., sent to a third party such as West/Comtech TCS/Bandwidth) and displayed correctly on the workstations. All calls where "no record found" occurs must be investigated.

8.4.3 Timing

As the project team looks at the calendar and determines the timing of the cutover, it should avoid holiday weeks, especially the Thanksgiving to Christmas to New Year's Day timeframe. Summer can be an issue due to vacations with possible implications for call-taker experience and expertise. The time of day to cut over is discussed above as a general good practice. Generally, begin the cutover on a weekday, starting mid-morning, preferably not on a Monday or Friday.

Before the cutover begins (see Chapter 7), there must be a freeze on additional enhancements and work. This ensures that the team is aiming at a stationary target.

NG9-1-1 network systems are complex enough without adding the chaos of evolving or radically changing requirements. The project team needs to know on the cutover date that all the functions and options have been tested, and that once completed, all the settings remain static so confidence can be high as live traffic is moved to the NG9-1-1 network and systems. Facility testing, layout, and labeling are addressed in Chapter 7; preliminary testing must be completed satisfactorily to make the go/no go decision on whether to proceed with the scheduled cutover date.

8.4.4 Training

Training has an enormous impact on the success of the cutover. If the people using the system are poorly trained or have forgotten the training, disaster looms; insufficient training can lead to frustration and the feeling that the system does not work like it is supposed to. That can create a feeling of dissatisfaction with the public safety organization.

Training the call takers and the PSAP managers must be done on a fully functional test system. The users need to be very familiar with the new NG9-1-1 user interface(s) and applications for such functions as call taking, transfer, text, map, CAD, and radio. These individuals need to be comfortable with the screen display and the options that are present to interact and operate the NG9-1-1 system and features. Training too far in advance may be counterproductive since memory fades quickly; unless the individuals apply the training soon after receiving it, it still may be necessary to refresh the information. Training as close to the cutover date as possible is the best option, but it can create some logistical issues for public safety and the people responsible for the trainers.

Setting up a training room with several workstations that are housed in a geographically convenient location is the optimum situation. If the geography makes traveling to the training an issue, setting up regional training rooms may be the best option. As a last resort, it may be possible to set up the new equipment at each live workstation where there is access to the fully functional test system and then to do remote training with hands-on experience.

8.4.5 Tabletop Cutover Exercise

A step-by-step walkthrough of the minute details of the cutover plan should be done with multiple iterations and an increasingly larger audience. This exposes weaknesses or gaps in the cutover plan. The team needs to review the logistics of having people available to make and receive test calls on cutover day. The team needs to review their own internal maintenance windows and those of the affected surrounding public safety agencies. This exercise will likely occur while system testing is under way. Confirmation of the database connectivity and the new e2 datalink connections will occur and be ready for live traffic. Secured authorized access to buildings, including each data center, must be arranged in advance as required by OSPs and/or the 9-1-1 SSP. Testing tools and systems including NOC personnel, monitoring systems, and ticketing systems must be included in training and readiness.

The cutover plan should clearly identify the individuals responsible for the following roles.

- Project manager;
- Technical lead;
- Testing lead;
- Data—GIS/ALI/MSAG Lead;
- OSP liaison;
- Design engineers NG9-1-1;
- Design engineers telco;
- Telco interface;
- Translations;
- Facilities;
- PSAP managers;
- Adjacent PSAP managers;
- OSP team leaders for the activity per detailed schedule;
- NOC managers and contacts;
- Data center managers and contacts;
- Security, logical, and physical contacts;
- ESInet POI location managers and contacts;
- Vendors on standby for all critical;
- Systems and components;
- Legal regulatory personnel on standby as needed.

8.5 Pressure-Test the Plan

Walk through the cutover plan with experts, and seek out people with experience in telecommunications and 9-1-1 conversion activities. The team should ask questions, perhaps using a devil's advocate approach, to ensure to the extent that possible the team has identified any loose ends or areas that need special attention. *Review the plan frequently with a critical eye.*

8.6 Communicate the Plan

The cutover plan should be reviewed in the weekly team meetings and in special meetings with people who are not part of the weekly meeting.

8.6.1 War Room Concept

All the participants with significant roles should be included in a "war room." This is a gathering point during the cutover to track progress, and the lead cutover planning team serves as the guiding force in problem-solving and making command decisions. The war room may be a physical space or a virtual space, depending on the participants. Ideally there will be at least two voice bridges and perhaps a video/screen share service.

8.6.2 Go/No-Go Decision

The decision to proceed to cutover or back out and reschedule (go/no go) will be made the afternoon before the cutover is scheduled. The team needs to review the latest test results and determine whether the functionality is sufficiently tested and ready for live traffic. If the test results show that the system is functional, and the other team leads see no issues that would negatively affect service, the "go" decision should be made. If there is any indication that any part of the network or system is not performing as specified, the cutover should be postponed. Other factors that may cause the cutover to be postponed would be severe weather, civil unrest, or some form of local catastrophe that could imply that large volumes of 9-1-1 traffic might occur in the cutover window. In that event, the cutover team will evaluate the practicality of cutting the NG9-1-1 system that specific day. The team should decide in advance who will have the final call to proceed or postpone the cutover on any day and have a rapid communications plan to alert all parties including commissions and executives with a need to know.

8.6.3 Back-Out Plan

In case of emergency the back-out plan should include the items below.

- Identify factors that could cause the cutover to be delayed;
- Prepare alternative dates in advance;
- Ensure that the existing E9-1-1 trunks are retained in place;
- Ensure that OSPs can change switch translations back;
- Ensure that the onsite technicians can reinstall the old workstations;
- Retest E911;
- Retest existing E9-1-1 trunks.
- Even if the cutover proceeds, the following steps are necessary:
- The existing E9-1-1 trunks need to be left in place for *n* days after the PSAP is cut. (Team or regulatory agencies determine the *n* number of days.)
- The E9-1-1 trunks should be disconnected *n* calendar days after the cutover date with the exception of the legacy SR interfaces for split exchange routing and possibly enterprise customers to be cut over later.

8.7 Execute the Plan

8.7.1 Execution

Final steps include the following:

- Live network verification;

- Workstation verification.

Vendors have been notified of the cutover and are on the bridge(s) as scheduled. The OSPs and PSAPs most likely will be cut over during a weekday from 9:00 a.m. to 3:30 p.m. local time. This accommodates most PSAPs since demand is historically the lowest during these times. It also increases the probability that sufficiently experienced personnel are available. In any case, project managers need to adapt to the needs or schedule of their PSAP.

8.7.2 Steps to Cut Over a PSAP

The following items are required to successfully cutover a PSAP.

- Change out half of the workstations (may vary, depending on traffic load strategy);
- Verify that new trunk groups are active;
- Reroute originating translations to the new ESInet trunk group(s);
- Place 9-1-1 test calls over new route configuration(s);
- Verify appropriate PSAP answers and verify location information;
- Determine the performance of the split exchange routing if applicable; test it before and validate it during cutover;
- Verify that transfers in and out are operational.

If a service interruption occurs, the following steps are necessary:

- Cease all work operations immediately.
- Notify local 9-1-1 SSP, agency executives, OSPs, and possibly commissions of outage details immediately.
- Provide a written report/call as needed.

8.8 Fix What's Broken or Unexpected

The break/fix phase begins the second the public safety agency accepts the system with live traffic. Now the issues in the system need to be triaged by designating them as: (1) service-affecting or (2) minor issues. The monitoring systems must be in place to alert the appropriate virtual or physical responsible NOC. The ticketing system must be in place to enter the problems identified and distribute actionable tickets to the technicians and vendors that can resolve the issue.

8.9 Review the Cutover

Post-cutover review should occur within a few days after the cutover. This is an autopsy of the project. The team should identify which best practices, problems, errors, and mistakes can be useful in the future.

8.10 Post-Cutover Legacy System Removal

After some predetermined timeframe, the legacy network should be disconnected and torn down and the legacy equipment removed. This timeframe may be discretionary or mandated by regulatory rule(s). Refer to individual state regulations for guidance. Also remember that it may not be possible to fully retire the legacy SRs. In point of fact, there may be more than a single legacy SR in a region or area to be cut over. The complexity cannot be underestimated. The actual cutover plan is detailed and extensive and has date/time stamps; the project manager marks off each step as complete before the next step takes place. It is important to practice timing the activities, so the schedule is not compromised. Any residual traffic or phases that must be scheduled for cutover should be well-documented along with lessons learned.

8.11 Cutover Strategy Checklist Example

The following is an OSP cutover checklist.

- The cutover strategy must be agreed upon by OSPs, data centers, and public safety agencies.
- The cutover plan validates that the network translations are operational for primary, alternate, and disaster situations to ensure that the proper response gets to the PSAP even if the network is overloaded.
- Cutover occurs OSP-by-OSP in the mutually agreeable OSP, data center, public safety agency and PSAP maintenance windows for 9-1-1 services. The process includes the following:
  - Optimization of physical and logical routes;
  - Outlining and confirming the circuit-ordering process;
  - Field testing resulting in cutover-ready transition;
  - Agreement on primary and alternate routing rules;
  - Agreement on split exchange routing and interfaces;
  - Translations for load-balance and routing in the carrier OSP network;
  - Sharing of OSP test numbers;
  - Addition of OSP test numbers into the NG9-1-1 database records;
  - SR trunking to data centers;
  - Originating rate center (NPA-NXX) to PSAP;
  - Verifying PSAPs before and after dispatch;
  - Making access trunks busy and verifying load balance;
  - Logging off workstations and verifying primary/alternate PSAP.
- Traffic goes live after all steps are complete.
- When all tests are successful, the first access traffic load is delivered to the appropriate PSAP/PSAPs in accordance with the deployment schedule.

8.12 MOP Description

The following is a sample checklist of methods and procedures for cutover:

- Setup and preparation: OSP trunks have all been installed and tested. 9-1-1, PSAP equipment, and data center hardware have been installed and tested. This is the procedure to cut over live traffic to the NG9-1-1 network and systems installed by the new 9-1-1 SSP.
- PSAP cutover: The plan is to cutover one PSAP at a time. Start at 10:00 a.m. on MM/DD/YY.

1. Put monitoring apps and Wireshark and other apps and systems in place to assist with alerting, troubleshooting, and ticketing. Ensure that the whole list of SS7 PCs and IP addresses are available along with access to the DNS servers.
2. Position personnel as required.
3. Set up the bridge and bring in each tester.
4. Test the existing network to existing PSAP to verify that works.
5. Change translations at the OSP switch.
6. Use the maintenance bridge to bring in necessary parties and trouble shoot.
7. Make test calls.
8. Verify that the primary and alternate load balance at ESRP is working.
9. Verify TTY on the wireline OSP lines.
10. Verify that voice quality is clear, with no echo, noise, or quality issues. Test and resolve before closing. As needed, employ systems to test for phase jitter or delay.
11. Document details.
12. Repeat steps 2–11 each OSP cutover.

8.12.1 Wireless Carriers

Wireless carriers should be cutover first. Wireless OSPs cut one at a time in designated order. The first OSP is asked to cut its traffic to the new system. Test calls are generated to ensure success. Once that is successful, the remaining OSPs are cutover live to the new system in the same way.

8.12.2 VoIP MPC/VPC Carriers

After all the wireless OSPs are cutover, the VoIP OSPs should be cutover next. VoIP and MPC/VPC OSPs are cutover in a predetermined order that will be published before the cut. The first VoIP/ MPC/VPC OSP will be asked to cut over their traffic to the new system. Test calls are generated to ensure a success. Once successful, the remaining OSPs will be cut live one at a time to the new system.

8.12.3 Wireline OSPs (ILEC, RLEC)

The legacy SR routes and interfaces will be tested first then the rest can follow.

Wireline ESNs for County Sheriff Office

ESN	*PSAP*	*Wireless/VoIP*	*Notes*
xxx	County sheriff	VoIP	
xxx	County sheriff	Wireless	
xxx	County sheriff	Wireline	

Once the ESNs are all cutover, the 9-1-1 SSP and public safety personnel will make test calls from predesignated test numbers to ensure correct call routing. The split exchanges need to be tested to ensure proper routing of calls to the correct PSAP(s). 9-1-1 SSP NOC will monitor calls and ensure proper call routing and delivery.

The police departments (PDs) should be cutover the same day as the sheriffs' offices.

Wireline ESNs for PD

ESN	*PSAP*	*Wireless/VoIP*	*Notes*
xxx	PD	VoIP	
xxx	PD	Wireless	
xxx	PD	Wireline	

Once the ESNs are all cutover live, the 9-1-1 SSP and public safety personnel will make test calls from predesignated test numbers to ensure correct call routing from both wireline and wireless calls. 9-1-1 SSP will monitor calls and ensure proper call routing and delivery.

Public safety personnel will make test calls from predesignated test numbers to ensure correct call routing for both wireline and wireless calls. The 9-1-1 SSP NOC will monitor calls and ensure proper call routing and delivery. The detailed end-to-end test plans are reflected on spread sheets with full routing details and NPA NXXs and test numbers and anticipated routes to show where a call may fail. Wireshark assists in identification of any points along the path where calls may fail. All parties involved in testing must be trained to use Wireshark and other tools. Vendors are to be on standby as tier-3 technical support to assist with any problem so testing may continue moving forward. Assuming that all the piecemeal testing is done as designated in Chapter 7, the actual cutover testing should simply be a repeat of what was accomplished long before cutover day.

8.13 Summary

There are many options for converting from E9-1-1 to NG9-1-1. Because of the complexity in the cutover each OSP MOP must be documented to include contact information. To begin the cutover the primary and backup plans are fully documented. The 9-1-1 SSP and agency lead project managers create and distribute the master plans to all parties for the tabletop exercise and the modified plans as they evolve. It is suggested that key responsible parties sign off on the plans and the MOP, officially designating its completeness and accuracy. Chapter 9 discusses ongoing monitoring and alerting, and the ongoing role of OA&M. Chapter 9's appendix includes a formatted MOP as a sample. A strong sense of accountability and teamwork is essential to the success of the cutover. Every community is different, and that uniqueness makes this a daunting project; careful planning leads to success.

CHAPTER 9

Operations Administration and Maintenance

9.1 Design for Success—Plan for Failure

OA&M includes the processes, activities, tools, and standards involved with operating, administering, managing, and maintaining any service, system, or network.

Include a budget for OA&M when a project is being planned. A rule of thumb is to allocate approximately 10% of the budget to this activity. NG9-1-1 networks must interoperate with older networks and equipment, and this interoperability is critical to successful emergency call completion. OA&M end-to-end is essential, especially when networks transition from E9-1-1 to NG9-1-1. Using standards based NG9-1-1 design supported by OA&M principles reduces the likelihood of failures. Yet in the real world, failures will occur. It is necessary to react in near real time to mitigate these failures: to minimize the number and impact of their occurrences; to document the failures; to identify their root causes; and, based on this information and analysis, to make the necessary changes, including, for example, changes in personnel, training, processes, network architecture, and monitoring systems. "That which switches with blinding electronic speed will fail with the same dispatch and with a total lack of grace." Anonymous. What is a failure? Operating failures may result from sudden disruptive events such as cable cuts, tornados, and earthquakes, and they may be the result of incremental changes in the operating environment, including demographic shifts that cause changes in traffic patterns that eventually overload network elements (NEs), or the result of temperature extremes in equipment rooms that may cause performance degradation. Events that impact service either abruptly or over time should trigger alerts to the personnel needed to respond. These events should be recorded in such a way that they can be analyzed, their root cause determined, and action taken to prevent failures in the future. When these events result in outages that directly impact a large number of customers, they may need to be reported to the appropriate overseeing government agency. Section 9.8 includes sample outages with best practices for monitoring, alerting, recording, and reporting. Section 9.10 describes the customers and regulatory agencies to which the 9-1-1 service providers are responsible; and it introduces the concept of RCA with continuous process improvement. Understanding the options and requirements in these areas will enable the 9-1-1 SSP to identify and select monitoring solutions, while the standards and regulatory bodies prepare to publish requirements.

9.1.1 Network Outage Types and Severities

9.1.1.1 Critical Outage—Severity 1

A critical outage impacts service to 9-1-1 callers. This has the highest impact since callers cannot reach 9-1-1. During critical outages, a conference bridge should be established by the 9-1-1 SSP to enable all parties to immediately address the issue until problem resolution. Immediate attention is given to the issue and response, and workaround or restoration should occur within one to two hours. At a minimum, a trouble ticket will be issued to NOCs. The 9-1-1 SSP is responsible for notifications to regulatory bodies for 9-1-1 service outages. See Section 9.4

9.1.1.2 Major Outage—Severity 2

A major outage refers to a major system module or function being inoperable with no workaround. The differences between critical and major are the level of the impact of the 9-1-1 calls. Immediate attention will be given to the issue, and response and workaround or restoration should occur within two to four hours. If there are major outages, a conference bridge should be established by the 9-1-1 SSP so the parties can address the issue and assure ongoing status is provided until problem resolution. At a minimum, a trouble ticket will be issued to NOCs if there is a trouble requiring attention.

9.1.1.3 Minor Outage—Severity 3

A minor outage generally includes loss of services and functions that do not affect call delivery. NOC tier I technician attention and/or vendor tier III support will be given the issue, with a response due within 24 hours. At a minimum, a trouble ticket will be issued if there is a problem requiring attention.

At the onset of contract negotiations with any OSP or vendor, it is important to identify the severity of each type of alarm and the response and restoration timeframes. This applies to the NG9-1-1 network, network components, and call paths. It applies to vendors that provide the equipment and/or are subcontracted for any work or component of the NG9-1-1 network, including but not limited to the data centers' NGCS and the ESInet.

9.1.1.4 Enhancements Other—Severity 4

A failure may be caused by something outside the scope of the original product or service description. Vendors and the 9-1-1 SSPs have contracts that include the exchange of trouble tickets between the 9-1-1 SSP and the OSPs and the 9-1-1 SSPs and their vendors. This process requires a mutually acceptable SLA definition for critical, major, and minor severity levels; also, a final category should be considered to include future enhancements, training, and documentation.

Sections 9.1.2–9.1.3.7 provide guidelines for negotiations.

9.1.2 Network Maintenance and Management

To improve coordination and information flow at critical times the parties should exchange information appropriate for the implementation of the network or service and to support the ongoing agreement. This would include maintenance contact numbers, network information such as circuit identifiers, information required to comply with law enforcement and other security agencies of the government, and a formal escalation process.

Each party must provide a 24-hour contact number for network issues. Escalation contacts and procedures will be provided in advance and readably available, Refer to Chapter 11.

9.1.3 Service Requirements

9.1.3.1 Network Monitoring and Mutual Alerting

The following assumes that each party has a NOC and real-time surveillance and alerting system(s) in place. Section 9.4 describes the roles of the NOCs.

The network management system (NMS) used for monitoring and alerting should facilitate rapid problem detection and alerting. If the 9-1-1 SSP believes that either data transport services or a portion of the data transport supporting services is compromised in any way, a 9-1-1 SSP NOC tier-I technician should contact each OSP, underlying vendor, and contracted provider or their own internal organization personnel and exchange trouble ticket information. If the external organizations' NOC personnel see an alert that they know or suspect will have an impact on the 9-1-1 service, it is incumbent on their NOC personnel to proactively contact and exchange trouble ticket information with the 9-1-1 SSP NOC.

9.1.3.2 Service Levels

The following service levels are included in NENA ESInet design for NG9-1-1 NENA 08-506, Version 1, December 14, 2011, and may or may not be inclusive or match the ultimate agreements between responsible parties. Thresholds for real-time alerting may be set at a level different than the service level agreed upon. An alert should be actionable. For example, if a threshold is broken for a couple of minutes it may or may not be significant enough for an alert.

Availability:

- NGCS routes to each PSAP: 99.99% availability;
- 9-1-1 SSP Data center or CO interconnect transport path: 99.99% availability.

Packet loss:

- ESInet transport: less than 1%;
- Data center NGCS to PSAPs: less than 2.5%.

Jitter:

- Less than 20 mS in the end jitter buffers.

Latency:

- End to end </= 150mS (ITU-TG.114).
- ESInet portion </= 35mS packet delay.

Automatic failover:

- The ESInet shall have automatic failover.
- The NGCS shall have automatic failover.
- The legacy gateway and security FEs shall have automatic failover.

Minimum bandwidth provisioned:

- This value is dependent on the size of the network footprint and the volume of calls predicted. Samples are provided as follows.
    - 1 Gbps on data center or CO facility interconnect;
- 100 Mbps on ingress from any Ethernet-connected OSP;
- 100 Mbps per PSAP.

9.1.3.3 Performance and Outage Reporting:

See Section 9.1.1. The following may seem contradictory, but realistically this is what often occurs.

Critical outages

A critical outage created by a total network failure such that no further call processing is possible will be immediately addressed by the 9-1-1 SSP with issue analysis and proposed resolution to be provided within one hour. The worst-case scenario goal is to have a workaround solution in progress with repairs completed within four hours for any facility or network failure creating or causing stoppage of customer call processing. In the event of a critical outage, a post-mortem meeting with RCA is required with results provided to the state commission and or FCC. The 9-1-1 SSP is responsible for calling and managing the RCA meetings. A technical representative from each company involved will actively participate in the analyses and action plans. The 9-1-1 SSP tier-II technical support manager will oversee data gathering, setting up the meeting and managing the results so the appropriate clients can be given feedback in a timely manner. If the event results in service outages, the "clients" for the final report may include: PSAP managers impacted, 9-1-1 SSP participants in the process, and FCC and or state 9-1-1 regulators. Financial penalties may be involved based on SLA agreements.

9.1.3.4 Design and Engineering

Design criteria and meetings

Semiannual design review and planning meetings should take place between OSPs and or vendor managers and the chief technical officer (CTO) of 9-1-1 SSP.

Additional participants may include personnel responsible for the following types of work: ESInet expansion and growth planning and data center requirements to cover performance and/or to address expansion for agencies that plan to join the NG9-1-1 footprint of the 9-1-1 SSP.

The 9-1-1 network diagrams, descriptions, and OSP test plans will be shared with the data transport and bandwidth providers after drafts are prepared.

OSPs will provide information reasonably requested by the 9-1-1 SSP for the plan modifications, as well as information required to maintain compliance with any regulatory rules and orders.

Each state and the FCC have their share of required reports and measurements, which may include changes to network designs and review of calls handled by each agency.

Diversity and redundancy

Power in the transport system facilities will have, where possible, commercial, generator and battery systems as needed. In addition, please note the following:

- Power diagrams should be reviewed and provided upon request.
- Power will be monitored in the NOCs, and any alarmed condition should be shared with the 9-1-1 SSP for alerting ticketing and reporting.
- Fire-suppression plans should be reviewed, shared, and maintained as needed.
- Environmental (HVAC) plans should be shared and maintained as needed.
- The network transport provided must comply with the specified network test plan as approved by the commission where required.

MOPs

Unless there are exigent circumstances, a MOP[1] process will be used for any planned activity to be initiated by either party and or their vendor who may need to enter the space logically or physically. The parties will provide each other as much notice as reasonably possible for any changes that require access or direct participation in the project. Vendor names and details of the potential impact on service, suitable time windows for change, and secure access as needed, either physical or logical access, will be detailed and agreed upon before any work will proceed. Regular planning meetings will be scheduled for any long-term project, such as the addition of a new client or interface into the network.

1. See Appendix A for a MOP sample that may be modified for local use.

Planned maintenance

The OSP or contracted vendors will give the 9-1-1 SSP a minimum of 14 business days' notice before any major data transport network work is required that may affect data paths. A MOP will be prepared and signed as required.

Maintenance windows

Maintenance windows will be negotiated. Most OSPs only permit planned maintenance from 10:00 p.m. to 5:00 a.m., and they do not allow planned cutovers upgrades or maintenance during peak seasons and other times. That includes blocking all planned work from Thanksgiving through News Year's Day, because staffing is the lowest during that time. Local and long-distance calling peak days fall into these timeframes. For 9-1-1 SSPs, the peak times and days of the week can be very different. Most 9-1-1 cutover work starts around 10:00 a.m. and ends by 5:00 p.m. during the day, Tuesdays through Thursdays. This is highly dependent on the environment. Where college campuses are involved, Wednesday, Saturday, and Sunday evenings may be peak times when 9-1-1 demand is highest. For others, there are seasonal times where 9-1-1 may be at its peak, such as in hurricane-prone areas, tornado and flooding areas, or where major events are scheduled including concerts, political rallies, and conventions. Maintenance windows for 9-1-1 should be addressed during contract negotiations. Exceptions may occur , and these must be negotiated and covered in a MOP.

9.1.3.5 Speeds

Actual data speeds may vary and are not always guaranteed. Many factors affect speed including: the number of devices using a single connection, the size and frequency of data to be transmitted, and the effectiveness or efficiency of CPE, network traffic. Customers who purchase 100-Mbps service(s) are provided data transport services at speeds of up to 100 Mbps, not continuous 100 Mbps. Customers who purchase 1-Gbps service(s) are provided data transport services at speeds of up to 1 Gbps; similarly, these are peak speeds. In most agreements, the transport provider will use commercially reasonable efforts to provide and maintain data transport speeds as close to the specified service(s) speed as reasonably possible.

9.1.3.6 SLA

In some RFPs or contracts there are monthly SLAs with penalties for financial reimbursement if the parameters are not met. Agreement ahead of time is essential to determine the level of severity and the penalties for not meeting them. All NG9-1-1 footprints for are different. All ESInets are different. Refer to Chapter 11.

Credit allowances (sample for ESInet)

SLAs often carry penalties that are a part of contract negotiations or an internal agreement between departments handling the ESInet within a company. For example, the ESInet provider will allow a pro-rata credit against future payment of the net monthly recurring charge (exclusive of nonrecurring charges, other one-time charges, measured charges, regulatory fees and surcharges, taxes, and other governmental and quasigovernmental fees) for a service interruption, except as specified below or as may otherwise be legally required (credit). For the purposes of calculating a credit allowance, the service-interruption period begins when the customer reports an interruption in the portion of the service to the ESInet provider, a trouble ticket is opened, and/or the service is released to the ESInet provider for testing and repair. The service interruption ends when the affected portion of the service has been restored, and/or the ESInet provider has closed the trouble ticket. Service-interruption time does not include interruptions of less than 30 minutes' duration. Suggested credits (length of service interruption per amount of credit) are listed as follows:

- At least 30 minutes but less than three hours yields a credit of one-eighth of a day;
- At least three hours but less than six hours yields a credit of one-quarter of a day;
- At least six hours but less than nine hours yields a credit of two-fifths of a day;
- At least nine hours but less than 12 hours yields a credit of one-half of a day;
- At least 12 hours but less than 15 hours yields a credit of four-fifths of a day;
- At least 15 hours and up to and including 24 hours yields a credit of one full day.

9.1.3.7 NENA Standards and Regulatory Requirements

Many of the items referenced were derived from NENA standards, which evolve over time. The authors have been part of negotiations to use the NENA guidelines and adjust for contracts and then again adjust monitoring settings in the systems to alert the proper parties and set changes in motion that make good sense.

9.2 Monitoring and Alerting

Monitoring and alerting systems are a key part OA&M for NG9-1-1 service. Each monitoring and alerting system makes it possible to detect, address, and correct failure conditions and impending failure conditions in near real time. Since 9-1-1 standards for monitoring and alerting systems as they pertain to NG9-1-1 service are not fully defined, many emergency services RFPs do not specify detailed requirements for a monitoring and alerting solution to be integrated with the NG9-1-1 network from the start of the project. However, the absence of complete standards and requirements does not mean that this vital part of NG9-1-1 service should be neglected or postponed. Lives depend on their inclusion into the plans. This chapter opens up the topic with the hope that the reader will learn from what has been done and see what needs to be accomplished. There is plenty of room for vendors, engineers, and systems developers to make a difference when they build to RFP requirements.

Monitoring for 9-1-1 has historically been the responsibility of the OSPs. These entities own and manage nearly all the legacy SRs. With NG9-1-1, the monitoring and alerting responsibilities are split between OSPs with access to 9-1-1 and the 9-1-1 SSPs responsible for the NG9-1-1 network. The OSP handles call delivery to the NG9-1-1 network. The 9-1-1 SSP through the NGCS handles: querying the location database to discover the physical location of the caller; routing the call based on the caller's location; and creating the connection to the PSAPs. The NGCS; ESInet, including all routers and switches, legacy gateways, signaling interfaces, and security FEs; and BCFs, must be monitored. Ideally, power, environmental, physical and logical security alarms, and POI monitoring and alerting are included.

Each OSP is responsible for bringing emergency calls into the emergency services network. The 9-1-1 SSP along with its SI is responsible for the routing function that the legacy SR provides in the PSTN. Thus, emergency service becomes the primary responsibility of the 9-1-1 SSP that now manages multiple entities: OSPs for connectivity, vendors, legal regulatory applications, their staff and consultants. Despite this distribution of responsibilities, the government regulatory agencies and the PSAPs require that two types of organizations be responsible for monitoring, maintaining and reporting on the end-to-end quality of the emergency service, the OSPs and the 9-1-1 SSPs. See Figure 9.1.

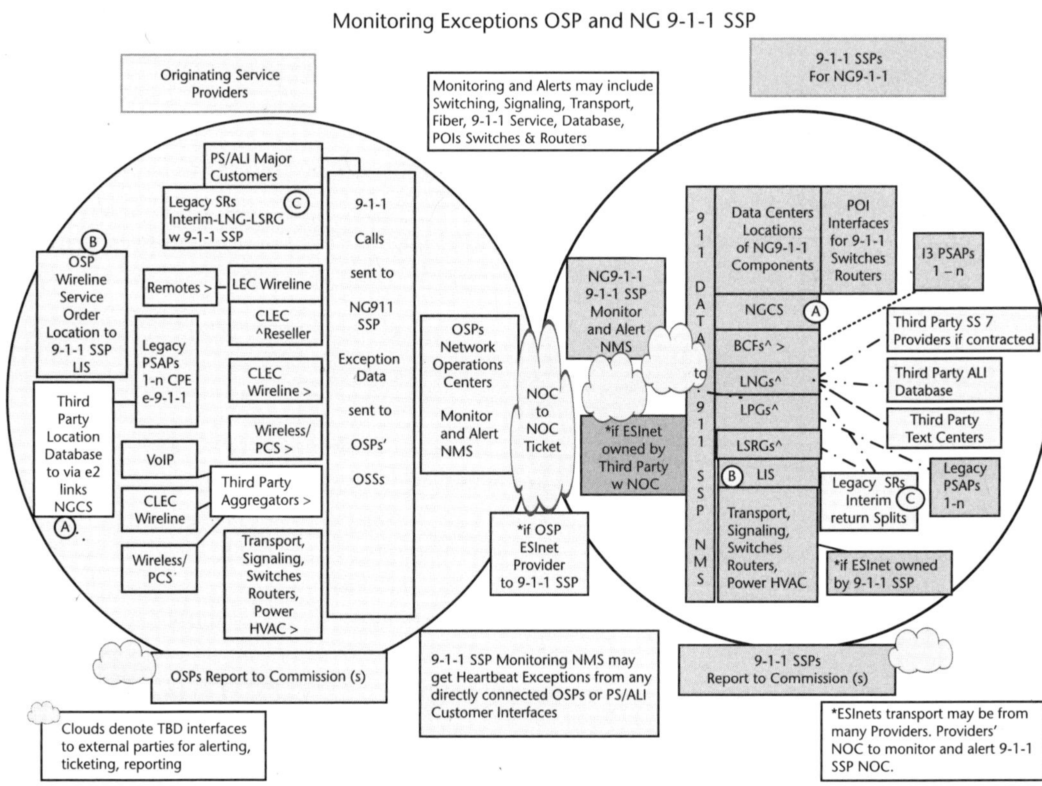

Figure 9.1 OSP and NG 9-1-1-SSP.

9.2.1 Where to Monitor

The monitoring, alerting, recording, and reporting functions need to cover the three areas of operation of the 9-1-1 SSP: the ESInet or physical IP network; the data centers or COs where the NGCS are located; and the PSAPs where the emergency call-takers are located. These areas are illustrated logically though not physically in Figures 9.2 and 9.3 and described in greater detail in Chapter 3. The best practices for monitoring and alerting in each of these three areas are derived from many sources, including the experiences of implementers, user teams, technical references, and specifications with examples provided in Section 9.5.

Monitoring the ESInet and the NGCS together with the managed IP networks and data centers must meet the same type of standards as the legacy PSTN-based network monitoring systems. The legacy networks use circuit-switched TDM networks to carry emergency calls, and the surveillance systems used on these networks are tailored to that technology. Next-generation networks are packet-switched and use managed IP-based networks, NEs, data switches, and routers, as their underlying mechanism for carrying calls. Managed IP networks have their own surveillance systems and best practices.[2] The monitoring systems must enable the alerting, recording, and reporting necessary to keep a lifesaving service available and efficient. The first level of monitoring starts with the network NEs. The

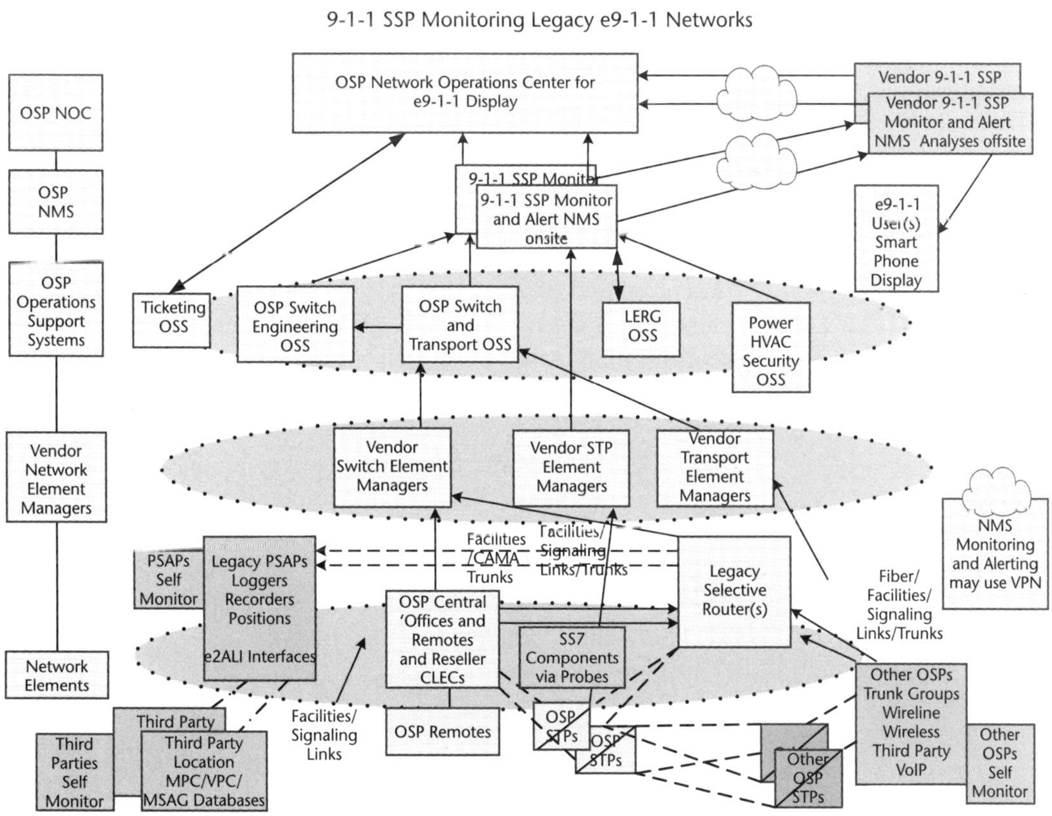

Figure 9.2 Logical monitoring for e9-1-1 networks.

2. See IETF, "An Overview of the IETF Network Management Standards," https://tools.ietf.org/html/rfc6632.

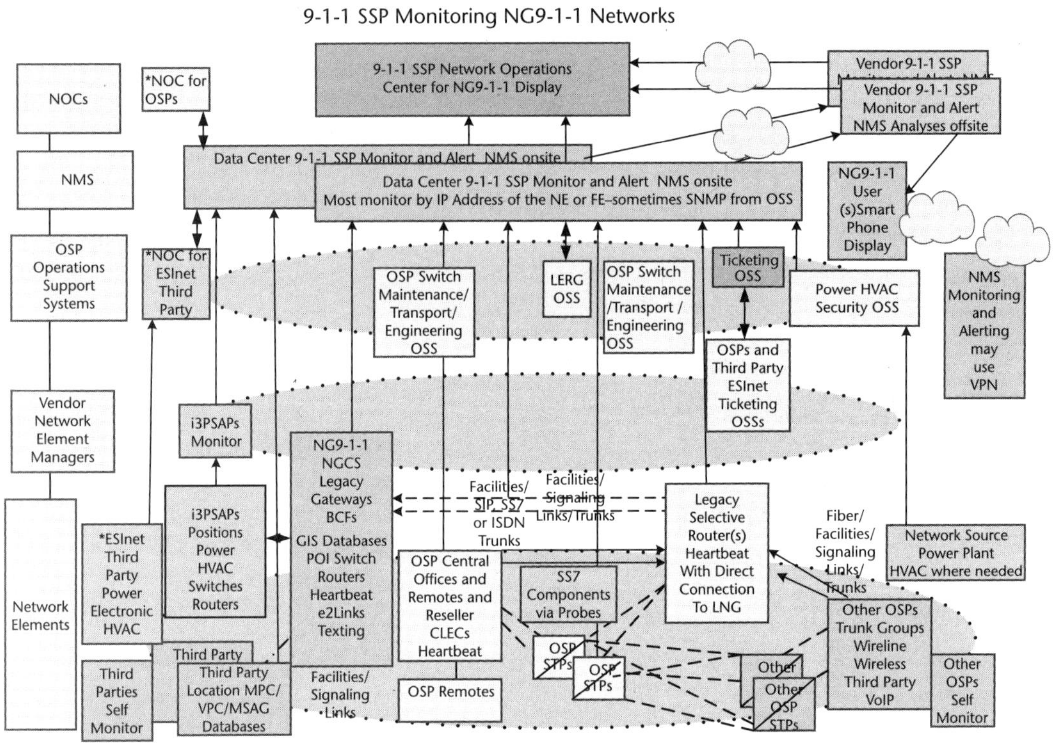

Figure 9.3 Logical monitoring for NG9-1-1 networks.

network element management systems (NEMSs) feed the next layer of monitoring systems called the network management system (NMS). The NMS uses the relationships of the NEs to build an end-to-end map. The last layer of monitoring leads to end-to-end network management (NM) with exceptions sent to the NOC level where human beings are presented with the alert including exception data, possible root causes and recommendations for next steps. NOC attributes are described in Section 9.4. Levels of action may include: tier-I dispatch, and or escalation to tier-II technical support or tier-III vendors for support in resolving any issue before it becomes service impacting. Everyone whose involvement is needed to resolve the problems in real time based on severity and population served by the NEs impacted by the event should receive alerts. Recipients thus may include database personnel, traffic engineers, and managers responsible for customer notification and/or commission notification. PSAP managers may need to be notified to implement staffing changes, open backup PSAP locations, or accept calls from failing or overloaded PSAPs with 9-1-1 calls that cannot complete for any reason.

9.2.2 What to Monitor

The call-processing NGCS, legacy interfaces, and ESInet must be monitored for any exceptions or incidents. Heartbeat monitoring on a very short duty cycle (i.e., the ability to send requests and receive responses from the product from a given spot on the network) should be included. Since an emergency services call may be carried across the boundaries of many companies, the use of a heartbeat monitor will help identify the cause of a lost connection or a deviation of call volumes from historical averages. The set of reportable exceptions associated with each product should be reviewed to ensure that platform-related and equipment-related events that may eventually lead to call processing failures will be recorded and available for analysis and remedial action before another failure occurs.

The database infrastructure—the LIS, ECRF and the ALI Links to the various internal and external location services providers—must be monitored for exceptions and failures. Without at least one ECRF, the calling party location information and the necessary instructions for routing each call will not be available. There are two groups that must see ECRF failures, the database managers and the tier-I technicians. Calls without location information or accurate location information may be routed to the wrong PSAPS. Chapter 3 defines the database infrastructure.

It is also important to perform real-time environmental monitoring of the sensors related to the HVAC systems, power equipment, generators, battery alarms, and security systems and fire-suppression systems. Failures in these systems can result in major outages.

9.3 Monitoring Data Centers and PSAPs

Currently, the 9-1-1 SSP for each NG9-1-1 network develops its monitoring requirements based on its experience and that of the organizations with which they contract. These requirements can be summarized as follows:

- The NM OSS must monitor the ESInet; data centers; and COs, including their signaling, switching, data functions, and infrastructure sensors.[3]
- The NM OSS must also monitor the PSAPs, including all their equipment, telecommunicator login status, and infrastructure sensors. Alerts based on the results of system monitoring must be sent to the appropriate organization. These alerts can be sent using text, email, trouble ticket generation, NOC OSS alert, or handheld NOC-like alert and display. The NM OSS must be able to provide critical near-real-time understandable exception information to all participating organizations including wireline, wireless, and VoIP OSPs, and database service providers.

9.4 The NOC

Where ticketing systems are in use, the NOCs will exchange ticket information when events begin and close them when they end. If no ticketing system exists, the information will be retained in a secured way (i.e., Excel spreadsheets) and be avail-

3. Infrastructure sensors may include power alerts from power feeding switching, signaling, and transport equipment, inverters, generators, and battery. Infrastructure may include failed environmental alerts such as temperature, door alarms, and others appropriate to a central office, data center, transport equipment in a remote POI, physical and logical security systems, and whatever is installed along the call path end-to-end that is deemed important to the emergency services.

able for review for at least one year after each event. Automatic interfaces between ticketing organizations are encouraged and in some cases may be required.

Each OSP has one or more NOCs that cover its whole footprint. Ideally each 9-1-1 SSP has its own NOC that receives exceptions from the many NOC OSPs that deliver 9-1-1 calls to them and exceptions from the underlying NEs for which they are responsible. The legal CAs between OSP NOCs and the 9-1-1 SSP NOCs includes language for joint problem resolution. The 9-1-1 SSP NOC is able to work with the appropriate OSPs' NOCs to prioritize restoration of emergency services. Within the 9-1-1 SSP area of responsibility, the 9-1-1 SSP NOC receives alerts and manages problems, using NOC OSS displays and tools such as conference bridges and other forms of communication to bring all problem solvers together rapidly. One method of bringing everyone together is using a handheld device alerting external users and off-duty personnel that a 9-1-1–related incident requires their support and possible action. To accomplish the distribution of information to users outside the NOC, clear roles and responsibilities must be established.

Figure 9.4 illustrates the sources from which the NG9-1-1 SSP users could receive alerts. For 9-1-1 SSPs these include the data centers or COs, the NGCS, legacy gateways and security systems, logging and recording systems, PSAPs, and the transport and signaling network that connects them. Alerts from all of these must be received and processed by the NMS of the 9-1-1 SSP. The methods and procedures that allow the 9-1-1 SSP or its designated responsible entity to see and act on these alerts are vital. Section 9.5 includes a sample list of 9-1-1 SSP alerts, their severity, and how they were handled. Figure 9.5 shows examples of user roles and responsibilities for handheld devices to receive notification of alerts—customized per 9-1-1 SSP.

9.4.1 NOC to NOC

Each organization with a NOC will exchange real-time communications numbers and contacts with escalation names, cell phone numbers and email addresses, organizations, and main center numbers.

9.5 Best Practices for NOCs and Alerting Criteria

The NMS should only alert NOC personnel when there are exceptions to thresholds. Incidents that trigger threshold exceptions may not always be understood from the first alerts. Because of the critical nature of 9-1-1, thresholds must be set differently for 9-1-1 exceptions than for other network services and network elements. A single failure may trigger a 9-1-1 alert. A single missing heartbeat may trigger a review of the facilities that are carrying traffic to and from the network as well as to the facilities that are carrying the alerts themselves.

Engineering standards for other services dictate that no more than one call in 100 in the busiest hour of the busiest season should fail across a 20-day time-consistent rolling-busy-hour average. It is ideal for engineers to receive an alert when a trend is developing that exceeds the engineering criteria. NOC alerts should be generated for 9-1-1 failures after a single failure occurs that impacts emergency call completion. The NOC 9-1-1 exception displays should show the network

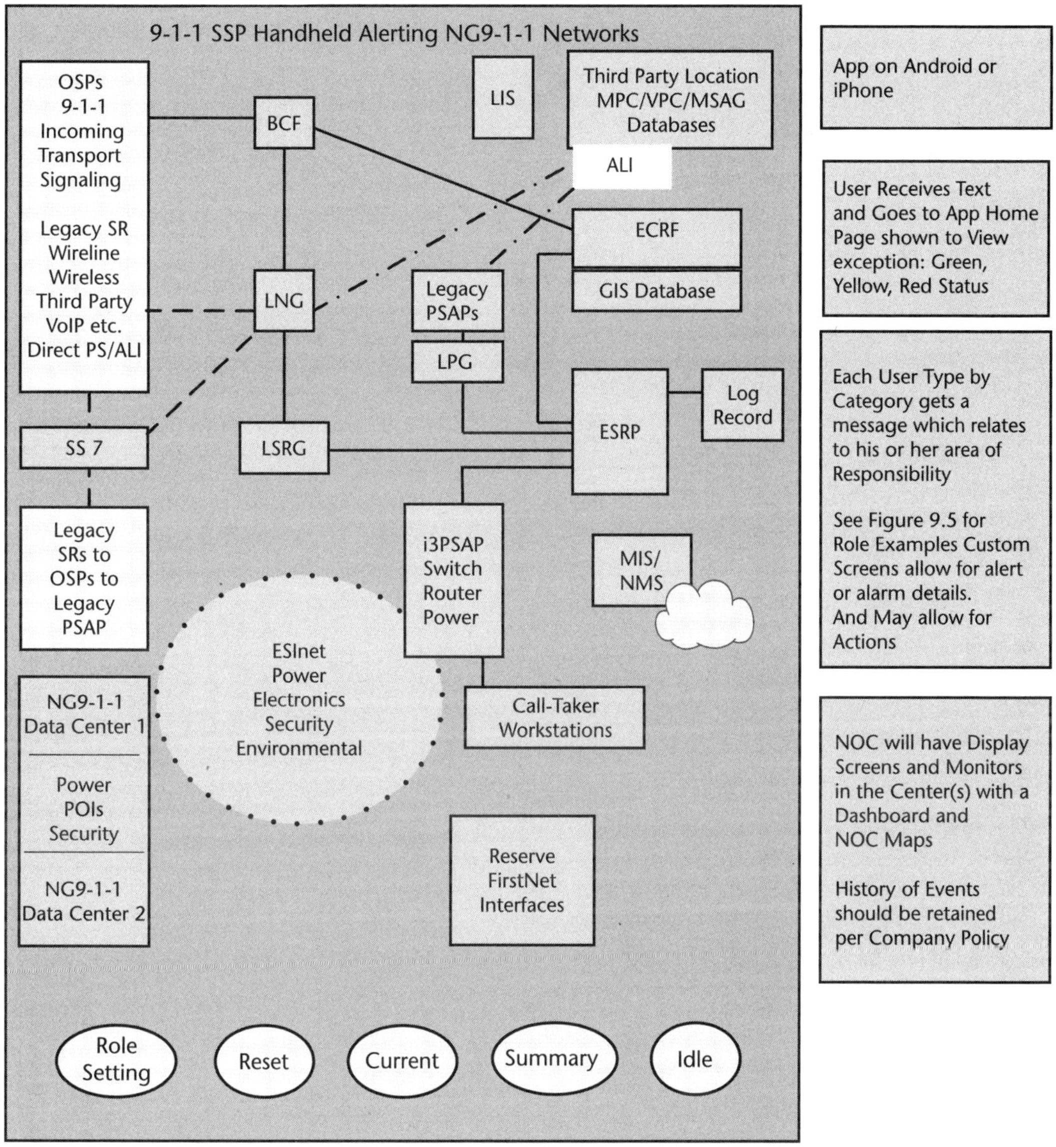

Figure 9.4 Handheld-alerting NG9-1-1 networks.

end-to-end. The system should map and highlight exceptions to the part of the network where a failure is detected. A NOC technician should be able to see the network image change color on a display and should be able to drill into the exception data to see the actual alarm or set of alarms that are mapped in real time. Among colors that are generally used, green means that all is clear; yellow means that simplex conditions exist that require attention before service is impacted; and red means that service is impacted.

The NOC module of the monitoring system should send escalation alerts to managers who need to be notified. Others with key roles ought to be alerted based on severity and the number of customers impacted. State or federal regulators may mandate notification in the case the number of impacted customers and or the duration of the outage exceeds their thresholds.

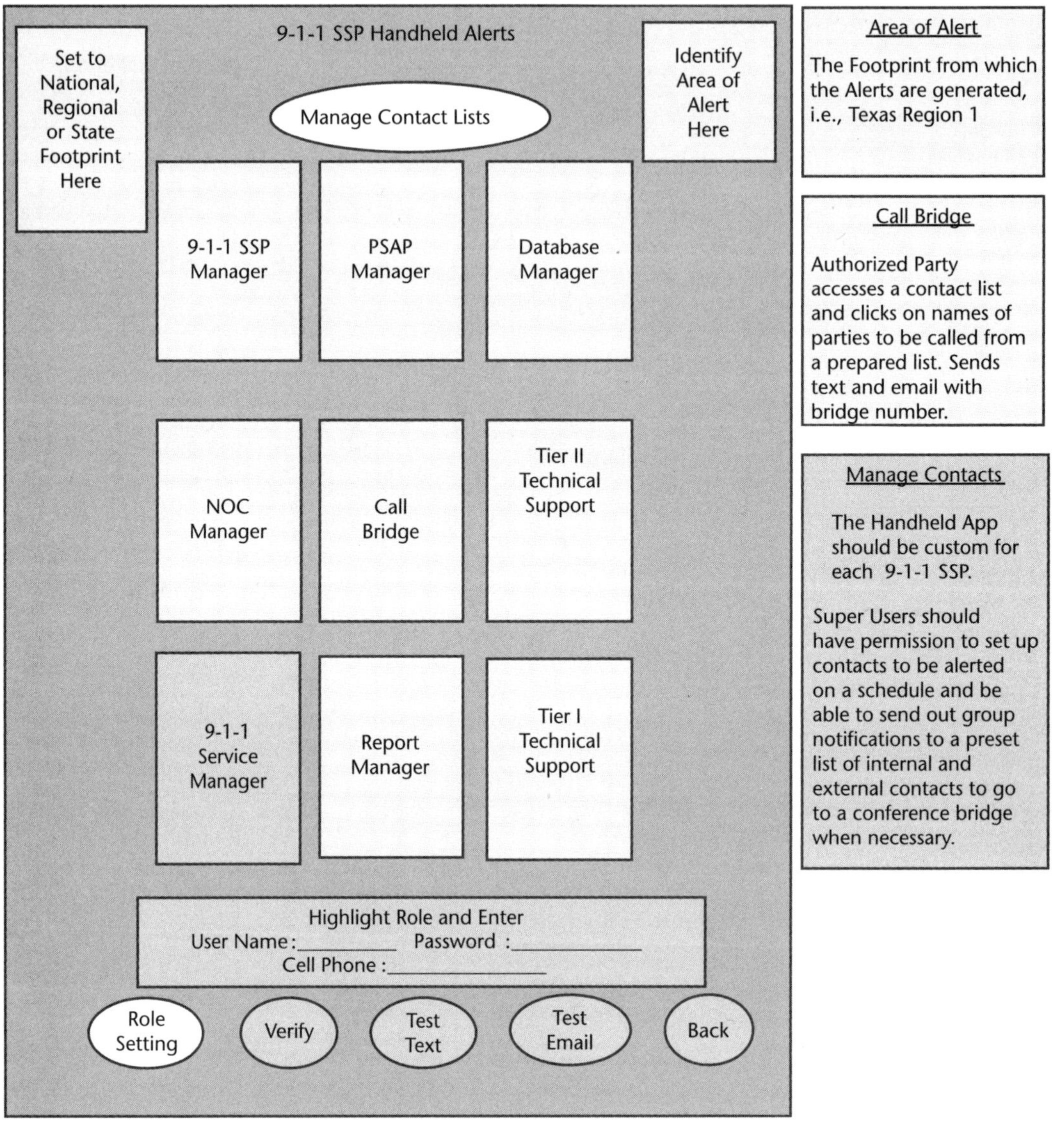

Figure 9.5 Handheld-alert select screen—roles and responsibilities.

The NOC should be able to generate trouble tickets automatically in response to alerts it receives. The NOC should be able to exchange trouble tickets with the various organizations whose participation may be needed to clear the trouble. These organizations could include tier-II technical support and managers, tier-III vendors, OSPs, and PSAP managers.

Database users should be alerted in the case of ALI link failures or database failures that impact call routing or the ability to locate the caller. While the NOC technician may be trying to resolve an individual ALI link problem, the database user must deal with the fallout of every single database failure. The causes of database failures may vary and may include slow service order processing, errors in the update process, improper translations, or ill-defined PSAP boundaries.

The network monitoring system should have user-definable decision trees in place to alert the right teams or individuals. This is an important collaborative process between the 9-1-1 SSP and the agencies they serve.

If there is no single NOC for the 9-1-1 service, then arrangements should be made to send automatic alerts to the proper parties with job function(s) based on their area of responsibility. A website can be set up and appropriately secured that will enable the responsible personnel to view the details and status of the alerts that they are receiving.

The NMS must detect and report significant deviations from normal patterns, such as a spike in the load or an overload. These nonfailure but abnormal patterns must be detected and logged and alerts generated. Abnormal patterns may require operational changes such as opening backup centers, shifting staff, or bringing in added resources. Such abnormal conditions may not immediately produce observable negative service impacts. The network's ability to generate alerts that trigger remedial actions may mask the occurrence. However, the fact that, in the moment, the incident did not impact service does not mean that nothing has failed. Recording the incident and its resolutions can provide a set of information that can be used to predict and prevent future incidents.

Ideally, the system provides pattern analyses from two sources: manual alerting thresholds set by the 9-1-1 SSP to personnel with designated roles and responsibilities, and dynamic alerting thresholds triggered by the intelligent NMS matching normal to abnormal trends per day and per hour from historical data.

9.6 Monitoring Data and Systems Requirements

The results of monitoring are used for several purposes: First, they serve to alert the appropriate staff members of conditions that require attention. Second, they must be recorded so that the aggregate set of event data can be analyzed for use in predicting and averting failures or degraded service. Third, they are used to generate reports to the relevant regulatory agencies that have jurisdiction over the OSP and/or 9-1-1 SSP as well as to inform other users of the 9-1-1 system including the public, the PSAP managers and staff, and the media. Based on experience and observation, the following requirements for monitoring and alerting systems or NMS should be met.

Since the 9-1-1 SSP may not be the owner of the ESInet's NMS or OSS, these two entities should have a contractual agreement in place ensuring that the interface between the 9-1-1 SSP's monitoring system and the ESInet's OSS will be able to carry information between the two systems in near real time and at all times.

9.7 Recording and Reporting

The federal and state governments have mandated outage reporting for NG9-1-1 systems but have not yet identified the sources from which the outage data should be taken. Commissions have not mandated cooperation between the various parties carrying the emergency call end to end. The absence of mandates must not prevent the involved parties from finding a way to obtain and report the information. The solutions that are developed today, and the experience gained deploying them will provide the basis upon which the regulatory agencies can create realistic rules of governance in the future. Section 9.7.1 discusses high-level reporting requirements based on the real-world experiences of 9-1-1 service providers.

9.7.1 High-Level Reporting Requirements

The system must report the number of customers at risk for a given reported event. Reports that include this information are mandated by the FCC and by some state regulatory agencies as well. The information provided must include the reasons that the report was issued. The 9-1-1 SSP must be aware of all local, state and federal reporting requirements. In some jurisdictions there are automated Reporting systems.

An expert system that learns the patterns of failure and typical root causes should be employed. This will enable the NOC personnel in the 9-1-1 SSP, over time, to predict and avoid failures by indicating corrective action be taken in advance of failures. The expert system should be able to provide the most likely causes of the problem, for example: power outage, fiber cut, loss of logging and recording devices, routers down, connections to carriers lost.

Displays must be provided that are useful to several classes or categories of users: database managers; PSAP managers; personnel responsible for making legal and regulatory reports; and the technical support personnel who must find and resolve the problem quickly.

The correct identification of the caller's location is a key part of the success of any emergency call system. The caller's location is used both to route the call to the correct PSAP and to direct the first responder to the right location to provide assistance. Location failures today are not usually reportable to a regulatory agency. They are the subject of FCC Notice of Proposed Rulemaking as mentioned in Chapter 11. The 9-1-1 SSP should record location-related errors and use the records of such failures to improve their processes.

The recorded information must be sorted and categorized for action. The appropriate action will be identified by expert systems that rely on accurate call flow information. Call-flow information is the building block upon which the expert system is built. The 9-1-1 SSP will need to provide network design documents with call flow information.

9.8 Real-World Incidents

The following set of incident reports are provided to give an idea of the type of incidents that are detected and recorded by a monitoring system in place in a real-time NG9-1-1 monitoring system today. The incident reports include fields that describe the nature of the incident and that classify it as major or minor, including descriptions of what caused the incident and what remedial action was taken. Other fields in the report indicate whether the incident was service-impacting, whether the FCC must be notified, and whether the state regulatory agency must be notified. The examples in Sections 9.8.1–9.8.7 are sorted in order of importance by the monitoring system.

9.8.1 Incident 1

Description: Data center failure—calls failed over via ESInet fiber ring 100% to the alternate data center.

Major: Yes.

Description: Commercial power explosion at data center 1.

Disposition: Alerting the NOC and action taken to restore working with commercial power company; backup power inverter and generator were available.

Minor: No.

Service-impacting: No.

Reason: ESInet design allowed for automatic failover of 100% of the load to data center 2.

FCC notification: Not required.

State/local agency notification: Informational.

9.8.2 Incident 2

Description: ESInet-planned outage.

Major: Yes.

Description: PSAPs to be impacted. Isolated PSAPs were notified of the plan and their backup PSAPs were prepared to handle the total workload.

Disposition: Outage extended well beyond planned approved MOP; real-time monitoring until completion.

Minor: No.

Service-impacting: No.

Reason: Design and routing Policy—the primary PSAPs were able to carry the overnight load. When traffic increased, the design of the routing function of the NG9-1-1 architecture allowed all other PSAPs in the footprint to accept the calls and transfer them back through radio or administrative lines for dispatch.

FCC notification: Not required.

State/local agency notification: Informational.

9.8.3 Incident 3

Description: Cable cut of fiber network string to PSAPs. The fiber string had multiple PSAPs on a single fiber without a ring, thus stranding the PSAPs. A fiber string is not an ideal design. There are not always multiple ways to physically reach a PSAP. Depending on size and the ability of other PSAPs on the network to handle the calls, this design was deemed acceptable for financial reasons by the 9-1-1 SSP and allowed by the state commission.

Major: Yes.

Description: PSAPs were notified of the unplanned outage and able to handle the calls on their administrative lines and radios to ensure proper dispatch.

Disposition: No calls were lost due to the use of radio dispatch and administrative lines.

Minor: No

Service-impacting: No.

Reason: Design and routing policy—the alternate PSAPs carried the load. When traffic increased, the design of the ESRP function allowed all PSAPs in the footprint to accept the calls and transfer them back through radio or administrative lines for dispatch.

FCC notification: FCC report not required. An FCC mandate of 2013 requires PSAPs to be notified any time a PSAP is not receiving calls or is isolated for any reason.

State/local agency notification: Required any time a PSAP is isolated even if service is being handled by an alternate PSAP. The PSAPs that were taking extra calls for the three inaccessible PSAPs were notified that the outage was in progress so they could staff accordingly.

9.8.4 Incident 4

Description: PSAP location switch or router down thus PSAP down.

Major: Yes.

Description: PSAP not receiving calls because it is cut off from the network by its own local networking equipment.

Disposition: PSAPs notified as each of these situations occurred so they could participate in restoration. Note that often there is someone working in the PSAP computer room who bumps or turns off a switch or router by accident. Switches and routers fail over time for various reasons and must be replaced. NOC monitoring of PSAPs should include monitoring of the switch and router equipment at each PSAP in addition to the PSAP CPE and workstations.

Minor: No

Service-impacting: No

Reason: Design and routing policy—the alternate PSAPs carried the load. When traffic picked up, the design of the routing function allowed all PSAPs in the footprint to accept the calls and transfer them back through radio or administrative lines for dispatch.

FCC notification: FCC report not required. An FCC mandate of 2013 requires PSAPs to be notified any time a PSAP is not receiving calls or isolated for any reason.

State/local agency notification: Required any time a PSAP is isolated even if service is being handled by an alternate PSAP. The PSAPs that were taking extra calls for the isolated PSAPs were notified that the outage was in progress so they could staff accordingly.

9.8.5 Incident 5

Description: PSAP position down—a nearly everyday occurrence and exception displayed even in small networks—sometimes due to an improper logout process.

Major: No.

Minor: Yes.

Reason: PSAPs have more than one position in service. Some PSAPs are small with only two or three positions, while others are larger; in some areas dozens of positions exist. The exception is important for PSAP managers and NOC personnel to know and rapidly respond to repair or reboot even when a single position is unavailable.

Service-impacting: No.

Reason: Same reasons as above; also note that other positions are often taking the calls except for peak periods where overflow is handled by the designated backup PSAP design routing rules.

FCC notification: FCC report not required.

State/local agency notification: Not required.

9.8.6 Incident 6

Description: ALI links down—data base missing.

Major: Yes, if all links failed; no, if some links are still up.

Minor: Yes, if some links are still up.

Service-impacting: Yes. The impact can vary.

Reason: Calls arrive but without the caller information; there is added risk that the call may not go to the right PSAP, because the right PSAP is unknown. If the person who responds cannot get information from the caller about their location, the situation may become life-threatening. Note that the landline database usually resides in the data centers and is no longer subject to ALI link database lookup. Unfortunately, many CLECs choose the VoIP model for database location services, which makes their database accuracy and associated reliability risks the same as those for wireless and VoIP callers.

FCC notification: FCC report not required. The FCC requires significant improvements in database accuracy over the next three to five years. See Chapter 11 for more details.

State/local agency notification: Not required.

9.8.7 Incident 7

Description: Access carrier OSP switch failure impacting 9-1-1 for all the customers in that CO.

Major: Yes. The OSP in question still uses its legacy SR before delivering the calls to the NG9-1-1 9-1-1 SSP. The OSP's legacy SR does not have an OSS interface to the current (NG9-1-1) 9-1-1 SSP that manages the network for

OA&M exceptions related to 9-1-1. The OSP has its own reporting criteria and responsibilities for 9-1-1 to the FCC and state. The 9-1-1 SSP's monitoring system could not detect this failure, because the OSP's CO was not directly connected to the ESInet. Many local citizens in the impacted area know their 9-1-1 PSAP dispatchers that they always talk to when they call and used their cell phones to contact the PSAP manager to say that their 9-1-1 calls were failing.

Minor: No.

Service-impacting: Yes.

Reason: Landline callers without cell phone or text access were unable to reach 9-1-1 or any number.

FCC notification: FCC report not required—did not meet 30,000-line FCC criteria for 30 minutes or more report requirements by the OSP.

State/local agency notification: The 9-1-1 SSP is not required to make reports for the OSP's access failures. It is unknown whether the OSP contacted the state 9-1-1 regulators, although it is doubtful that they did contact the regulators.

9.9 Reporting to PSAPs and Regulatory Agencies

There are groups of users who need to receive reports and/or notifications of incidents and events that impact NG9-1-1 service. This section describes the needs and expectations of these emergency services customers as they relate to recording and reporting.

Reports should be optimized to meet the needs of the group to whom they are provided. This ensures that the reported event is managed properly and that any mandatory reports are made in a timely manner. Fines and fees costing up to millions of dollars may be levied against 9-1-1 SSPs that fail to detect and report service-impacting events or that fail to provide timely updates to entities that have a need to know.

Legal actions may result when someone dies or loses significant property as the result of emergency services not being provided in a timely manner. Emergency events, including how they are managed or mismanaged, often make the news. Companies and public safety agencies lose credibility if they are perceived to be unreliable or unresponsive to the needs of the citizens for emergency services. The agency's staff loses credibility if the media cannot obtain current information from them. Loss of credibility translates to lost jobs, money, and investment. Accurate reporting and recording are a critical part of the emergency services operation. Sections 9.9.1–9.9.4 describe the four main classes of users of the emergency services system.

9.9.1 Designated 9-1-1 SSPs

The designated users of the 9-1-1 SSP that need to receive reports and/or notification include the regulatory agency, FCC, and any local or state authorities; the 9-1-1 SSP leadership; the PSAP personnel; and the media. The personnel in charge of public relations and media communications require notifications that include sufficient information to allow them to understand the impact of the incident on their operation and on the community.

Minimally the organization needs: the start and end times of the incident; the area and population impacted; the likely root cause of the problem; and restoration information.

Most companies with 9-1-1 SSP responsibilities want to retain control of the timing and messages to executives and to the regulatory agencies. The monitoring and reporting system should provide an initial report in a format that allows the responsible user to edit the partially populated fields; to gain internal approval; and to rapidly push the report to the pre-designated regulator(s). Reporting in a timely manner prevents fines and penalties from accruing. Initial reports may be annotated as "draft" with unknown root cause with estimated restoral times included. Within specified a time limit, a full RCA report is almost always required.

9.9.2 PSAP Managers

PSAP managers and PSAP personnel need to receive notification of all incidents that impact their own operation or the public whose interests they protect. There may be on-site issues a PSAP manager needs to know about, including positions out-of-service, overloads, power outages and electrical problems, or overheating in their equipment/computer room. If the PSAP receives massive numbers of calls due to a particular event, backup centers may need to be activated, staff may need to be added, and other PSAPs may need to be contacted for backup support. Alerts help the PSAP during extraordinary circumstances while restoration is progressing.

The types of incident that PSAP users must know about include: PSAP isolation from the network; the failure of a PSAP's onsite equipment or cabling; a PSAP environmental problem such as water on the floor, HVAC failures, over-heating, noxious fumes, and chemicals; the failure of the primary PSAP; when a PSAP is receiving a massive call overload; additional information gleaned from the data stream including sounds, text, or video.

9.9.3 Database Analyst

Database analyst users include GIS users who maintain databases; database managers who interface with external ALI database providers; and default database team leaders in designated PSAPs who get default calls when the location information does not arrive.

These users must know about ALI link failures as well as about incorrect or nonexistent location information. Dual-ALI link failures, for example, should be sent to the NOC and/or technicians as well as to the database manager so all failures can be grouped and managed accordingly.

Calls arriving without location are "at-risk" calls, possibly resulting in loss of life and/or property especially where the caller cannot identify location. This is a 24 by 7 by 365 concern, and a designated contact list must be maintained by the user community identifying who gets alerts on what schedule. Reasons why location information is not being received include: lateness of wireline service order information (SOI), inaccuracy of the VoIP and/or wireless provider's location data including VoIP-like CLECs that use third-party location services that fail to return the correct address or any address. Default PSAPs may need to be notified if a significant number of calls are failing toward them.

9.9.4 Technical Support

Personnel who provide technical support need: awareness of the incident; visual confirmation of the geographic area in which the incident is occurring; the ability to drill down to the alarm specifics, including start time; and the ability to change the status of an alarm, 24 by 7 by 365.

Technical support users are focused on the trouble ticket start and end time and resolving the trouble. This class of users should be able to begin restoration, using escalation policies, procedures, and contacts. The user team needs the ability to set up and edit the contact information for the target work center(s) or individual(s) who is/are responsible.

9.10 Root Cause Analysis

After any major incident or event impacting 9-1-1, the team working the incident needs to do a "post-mortem" or RCA review of the incident. The RCA is used in the case that an event or incident impacted a large number of customers; if it breaks a regulatory threshold, a formal RCA document may be mandatory to close out the commission's final report. The final reports can be due as quickly as 30 days after the reportable event.

The tier-II technical support organization that took control of the event is generally the best to lead the analyses meetings and prepare the final report, which is usually scrutinized by the leadership of the responsible entity. Sometimes a legal review is required to ensure the report complies with published law and accurately reflects what occurred. As part of an RCA, if the event was planned, the MOP review determines whether a step was missed or needs to change for any future event. For unplanned events or incidents, similar questions need to be answered.

The commission websites usually provide the key reporting criteria and where and how to handle reporting. Some monitoring and alerting systems provide a predefined list of items to be included in reports. After identifying the root cause, the next step is to identify the number of customers impacted, length of time start to finish, ticketing information, and what will be done to keep similar problems from occurring. That may include changes in processes and training or looking for vendors that meet the requirements of the network and customers.

9.11 Summary

This chapter discusses the need for a comprehensive plan for OA&M. We deal with the importance of monitoring, alerting, and reporting systems and incidents. Alerts are needed in real time to prevent or limit outages. The 9-1-1 SSP should analyze these events and patterns of events in retrospect to find ways to prevent the same problems from arising again. Analyses may result in plans for upgrades, process improvements, additional training, configuration, or product changes. Lessons learned can become best practices, and, if shared with others in the industry, they can lead to across-the-board changes that benefit everyone. To quote the FCC, "Finally, we believe the network monitoring component of the existing rule should cover not just the physical diversity of monitoring facilities, but also the proper prioritization of critical network alarms. For example, 9-1-1 service providers should have processes in place to send critical alarms to appropriate personnel whenever a substantial proportion of 9-1-1 calls are not being processed, comparing the number of calls originated to the number of ALI queries."[4]

Selected Bibliography

NENA-STA-010.2-2016 (originally 08-003) DSC Approval: 08/16/2016 PRC Approval: 08/29/2016 NENA Executive Board Approval: 09/10/2016 – i3 Design.

NENA ESInet Design for NG9-1-1 NENA 08-506, Version 1, December 14, 2011—Availability.

NENA NG9-1-1 Transition Plan Considerations NENA-INF-008.1 (previously NENA 77-501) NENA 75-001 Security Requirements, December 20, 2013.

FCC-14-186A1, POLICY STATEMENT AND NOTICE OF PROPOSED RULEMAKING, Adopted: November 21, 2014, IV. NOTICE OF PROPOSED RULEMAKING, A. Revisions to Rule 12.4, Paragraph 45, Page 20, Released: November 21, 2014.

FCC Rulings on Location Services, CSRIC, PSAP Notification, and NORs Reporting.

IETF Overview of the IETF Network Management Standards https://tools.ietf.org/html/rfc6632 NENA Working Group on Monitoring has not produced a Report as of December 31, 2018.

4. Document FCC-14-186A1, POLICY STATEMENT AND NOTICE OF PROPOSED RULEMAKING, Adopted: November 21, 2014, IV. NOTICE OF PROPOSED RULEMAKING, A. Revisions to Rule 12.4, Paragraph 45, Page 20, Released: November 21, 2014.

CHAPTER 10

Emerging Solutions

10.1 Overview

This chapter is a bit like the movie "Back to the Future"[1] in that we travel back to the vision for NG9-1-1 and FirstNet. That vision created demand for new services that set the stage to reach for the stars creating the wishlist, which turned into requirements and plans for emerging and futuristic public safety and emergency services. The authors see the NG9-1-1 network and systems as a foundational backbone for attaining the future, complementing the emergence of advancing access methodologies and technologies. The existing E9-1-1 network and technology has had many years of "force-fit" modifications and solutions to keep service running; we are now running out of time and a whole new end-to-end solution is required. The vision has always been SIP end-to-end while the wireline world is rapidly diminishing.

10.1.1 Demand for Emergency Services—Access

Never have we seen such high public safety demand as in 2020. We have raging western wildfires; we have extreme storms, derechos[2] and tornadoes through the south, southwest, and Midwest; we have simultaneous hurricanes creating something called the Fujiwhara effect[3] making landfall in the south while other hurricanes travel north along the east coast causing floods and wind damage. The first sizeable pandemic[4] since 1918 has hit the world and shows enough strength to

1. "Back to the Future" is a 1985 American science fiction film directed by Robert Zemeckis and written by Zemeckis and Bob Gale. It stars Michael J. Fox as teenager Marty McFly, who accidentally travels back in time to 1955. (*Source:* Wikipedia.)
2. Derecho (from Spanish: *derecho*, "straight" as in direction) is a widespread, long-lived, straight-line wind storm that is associated with a fast-moving group of severe thunderstorms known as a mesoscale convective system and potentially rivaling hurricanic and tornadic forces. (*Source:* Wikipedia.)
3. When two hurricanes collide, the phenomenon is called the Fujiwhara effect. If two cyclones pass within 900 miles of each other, they can start to orbit. If the two storms get to within 190 miles of each other, they'll collide or merge. This can turn two smaller storms into one giant one. (*Source: Business Insider,* August 21, 2020.)
4. The COVID-19 pandemic, also known as the coronavirus pandemic, is an ongoing pandemic of coronavirus disease 2019 (COVID-19) caused by severe acute respiratory syndrome coronavirus 2 (SARS-CoV-2).

keep us scrambling and needing emergency assistance through at least early 2021. We are experiencing cyberattacks on a massive scale nationally and internally. The demands on public safety for fire, police, and ambulance services that are traditionally required have not abated: highway accidents, air disasters, shootings, heart attacks, strokes, drownings, chokings, and you name it, all happening daily and the persons needing assistance are rich, poor, young, old, and diverse in ethnicity, religion, culture, and ability. Increasingly people are on the move: when they call, text, or signal they need help; they want someone to be there, offer immediate triage, and as needed, dispatch the best resources to the right location so lives and property can be saved. Approximately 80% of callers have wireless or VoIP access services. They hold cameras in their hands and they text, much more than they make voice phone calls. They use social media, messenger, and apps even more than they call. With the pandemic, conference calls, once mostly the domain of the business community, became the prime means of teaching in lieu of in-person schooling. Visits with family have converted to Zoom, Google Duo, FaceTime, and other applications. Who has not attended a virtual party, concert, or cooking demo; participated in a dance class; had a telemedicine call or business meeting; or watched TV via their phone or small-screen technology? The amount of bandwidth that is being reallocated and augmented to meet demands is staggering. Emergency services need to keep pace with the technology in the hands of consumers. NG9-1-1 is the core for the public safety community to interface with new and now common sophisticated and evolving technology.

At the same time, the demand for emergency services is exploding, as natural and man-made disasters loom: PSAPs are flooded; others are in fire zones requiring evacuation; and the first responders who always had a backup plan in case they needed more manpower are now needing backup support many more miles from their primary PSAPs. Those who have deployed NG9-1-1 are seeing the immediate benefits of diverse access, a resilient ESInet, dual data centers housing shared redundant NGCSs and associated equipment more than 50 miles apart, and the ability to get calls with improved location technology. Those public safety agencies that have not been able to make the move to NG9-1-1 are indeed working harder and struggling to both handle emergencies and manage the demand in their own homes and communities.

10.1.2 Providing Emergency Services—Egress

10.1.2.1 FirstNet

The FirstNet network is an emerging dedicated, nationwide wireless broadband platform that enables high-speed data exchange among first responders. In 2012, Congress authorized its development and allocated both $7 billion in funding and the 700-MHz spectrum for its use.

The disease was first identified in December 2019 in Wuhan, China. The World Health Organization declared the outbreak a public health emergency of international concern on 30 January 2020 and a pandemic on 11 March. As of 13 September 2020, more than 28.7 million cases have been reported in more than 188 countries and territories, resulting in more than 920,000 deaths; more than 19.4 million people have recovered. (*Source:* https://en.wikipedia.org/wiki/COVID-19_pandemic.)

"The FirstNet mission is to deploy, operate, maintain, and improve the first high-speed, nationwide wireless broadband network dedicated to public safety. This reliable, highly secure, interoperable, and innovative public safety communications platform will bring 21st century tools to public safety agencies and first responders, allowing them to get more information quickly and helping them to make faster and better decisions. Firstnet.com." FirstNet is being deployed across North America. NG9-1-1 must connect to and fully interoperate with FirstNet to get the full benefits of the end-to-end vision for emergency services. FirstNet was envisioned for the first responders on site to be able to communicate with each other. It is important to reflect on the terrorist attacks on September 11, 2001, or 9/11 when hundreds of first responders from so many agencies were on the scene at the World Trade Center towers. The towers were starting to fail; the ability to coordinate the response and have seamless communications with all parties on the ground was impaired. Bandwidth was insufficient for the overloaded wireless networks, and disparate radio systems and spectrum meant that first responders were left without adequate means to notify or coordinate with each other. Not only were citizens' lives lost, but also we lost hundreds of brave first responders. It took years of legislative and funding efforts, but finally FirstNet contracts were awarded. It took vision and creative approaches in technology to clear spectrum and plan for the technology to prioritize public safety needs in the worst case disasters.

"Emergency response agencies long have recognized their communications systems lag far behind the capabilities of many consumer technologies. In 2006, the U.S. Department of Transportation launched a study to explore how current 911 services could be enhanced to support emergency communications in wireless-enabled environments. The study validated requirements for a network that will enable citizens not only to call 9-1-1 centers or PSAP by voice, but also to send text, images, and video from their mobile and networked devices. At the same time, PSAPs will be able to receive data from sophisticated sensors, wireless-enabled vehicles, second tier responders, and other PSAPs and agencies." NG9-1-1 is the network that will make this possible (https://www.dhs.gov/science-and-technology/news/2015/07/24/responder-news-overview-firstnet-and-ng9-1-1).

10.2 FirstNet Vision and Visionaries

One of the leading spokespersons and visionaries for FirstNet was Jeff Cohen, chief counsel for law and policy APCO wrote, "FirstNet is Almost Here."

"A united public safety community achieved great success this past February with enactment of the public safety broadband legislation. In all significant respects, public safety received the spectrum, funding and nationwide governance structure it lobbied for and which all agreed were essential to success. With the reallocation of the D Block for public safety broadband, public safety has the spectrum capacity it needs to address large scale energies, and to foster the public/private partnerships essential for network deployment and sustainment. As for the funding piece $7 billion as I've said before, $7 billion is a LOT of money but not also a lot of money, meaning it is a terrific win for public safety but expectations must be kept in check. So that leaves us with the nationwide governance piece, which in many ways is the most important key to success." Cohen goes on to make the points about the lack of focus when local and state organizations try to govern deployment of such an undertaking, words that certainly were prophetic when it came to the meager progress toward NG9-1-1 without oversight, funding, and a governance plan.

Resuming key points of his address, "We will vastly improve upon the way public safety networks have been deployed in the past. FirstNet will ensure a sustainable network design that includes maintenance and coordinated upgrades. It will act quickly and avoid the multitude and variety of procurement and approval processes." Skipping forward, the remaining statements dovetail with the goals of NG9-1-1. "FirstNet will ensure a nationwide level of interoperability-so important for mutual aid and truly a matter for homeland security and when faced with the next natural or manmade disaster. FirstNet will be responsible for ensuring as secure a network as possible-especially vital due to the public safety of this network and increasing cyber-related threats. A new highly game changing technology-LTE[5]- is suited to an efficient national architecture and yet by its very nature enables local customization (prioritization access)." Some of the university and private partners/leaders supporting tests and trials for NG9-1-1 are working on interoperability with FirstNet and the Department of Homeland Security (DHS). Public safety agencies are members of NENA and APCO. There is a constant push and pull for resources to support both initiatives even when funded. We advise those planning NG9-1-1 networks to fully understand the status of FirstNet in their area and be prepared to add interface testing once both networks are launched to optimize the end-to-end service goals.

10.2.1 Interface

So much is still on the drawing board and so much is yet to be completed. The fact is that there are major differences in funding and in the management of NG9-1-1 and FirstNet. FirstNet was given a national priority with funding and a contract award to a single provider, AT&T. There is no national mandate for NG9-1-1, and funding has been inadequate to nonexistent. Sadly due to massive 2020 protests and civil unrest, there has never been a greater need for both fully funded integrated 21st century solutions than right now. The deployment of 5G wireless networks can provide that bandwidth to help alleviate most wireless bottlenecks. Aging access infrastructure technology for landline services must be addressed, even though wireline OSPs are not launching new services applications. We must find the limitations and remove them, and in part that is what has driven our desire to write this book and outline some of the steps to progress. Let us proceed to look at how we arrived here and what is urgently needed. We look "back" and give homage to the past and credit to the visionaries and organizations moving forward "to the future."

5. Long-term evolution (LTE) is a standard for wireless broadband communication for mobile devices and data terminals, based on the GSM/EDGE and UMTS/HSPA technologies. It increases the capacity and speed using a different radio interface together with core network improvements. LTE is the upgrade path for carriers with both GSM/UMTS networks and CDMA2000 networks. The different LTE frequencies and bands used in different countries mean that only multiband phones are able to use LTE in all countries where it is supported

10.3 NG9-1-1 Vision and Visionaries

See "Next Generation 9-1-1 (NG9-1-1) System Initiative, Human Machine Interface Display Design Document, Washington, D.C., January 8, 2008, Version 1. "The purpose of this document is to describe the design for the human machine-interface (HMI) display for the NG9-1-1) System (or 'system of systems') based on the initial Tier 1 requirements identified for the NG9-1-1 System. The ubiquitous access to 9-1-1 emergency services needed in today's world of evolving technology is driving the need to introduce a broader array of interconnected networks that would comprehensively support emergency services. The U.S. DOT is leading the effort to introduce the NG9-1-1 System, which is designed to improve management of emergency services throughout all aspects of operations, including public access to the emergency services; facilitation of services; and delivery of emergency information to public safety answering points (PSAP), emergency call centers (ECC), and first responders. USDOT views the NG9-1-1 System as a necessary transition to enable the general public to make a 9-1-1 'call' from any wired, wireless, or Internet Protocol (IP)-based device, and allow the emergency services community to take advantage of enhanced call delivery and advanced functional and operational capabilities through new internetworking technologies based on open standards. By enabling access to 9-1-1 services through virtually any communications device, the NG9-1-1 System provides a more direct ability to request help or share critical data from any location with emergency services providers. In addition, call takers at the PSAPs will be able to transfer emergency calls to another PSAP and forward the location and other critical data, such as text messages, images, and video, with the call. The HMI within the NG9-1-1 PSAPs will assist in consolidating and presenting emergency information received via IP, cellular, telematics, sensor, and public switched telephone network (PSTN) devices to the call taker in an efficient end-user format. It will provide call takers with an interface to manage access to 9-1-1 requests for services received through virtually any communications device, provide a more direct ability to share critical data with emergency services providers from any location, transfer emergency calls to another PSAP, and forward the location and other critical data with the call. The HMI is envisioned to improve call takers' efficiency and reduce time in responding to emergency calls."

Looking back we realize the vision of the new services being integrated into NG9-1-1 is far from being accomplished. We give credit to NENA and key visionaries.

Laurie Flaherty was a driving force for USDOT funding for the 2006–2007 proof of concept (POC) efforts that led to the first laboratory trial of the NG9-1-1 concept at Texas A&M. Equipment remained in place, and that provided a backbone for the NG9-1-1 lab used in ICE events and significant testing of practical solutions.

SIP signaling is the backbone of the NG9-1-1 core network and the pathway to end-to-end flow of emergency access to egress. RTP signaling allowed voice to be added to the other means of signaling across IP networks and not be delayed or choppy. The software development for components of the NG9-1-1 network were developed in large part at Columbia University including efforts by graduate students. The SIP Forum[6] is instrumental in contributions supporting NG9-1-1. Some manufacturers and visionaries of aging 9-1-1 networks began their own development, according to the emerging draft NENA standards developed by the NENA committees, under the direction of Brian Fontes, NENA's CEO.

Brian Rosen had a strong vision for the emerging NG9-1-1 design. Between 2006 and 2009, Brian was senior director at Neustar, where he was the subject matter expert on 9-1-1 and was involved in many of Neustar's new product development activities. Brian was chair of the long-term development working group at NENA, which was writing the technical standards for NG9-1-1. He was a significant contributor to the Internet Engineering Task Force (IETF)[7] standards efforts, including those for location and emergency services. Location services remain a vital concern for all emergency services. When asked to explain his vision for NG9-1-1, Rosen came to the Illinois Institute of Technology (IIT) in Wheaton, Illinois, long before standards were approved, and did a short whiteboard presentation of the i3 vision. His drawing was a total transformation from E9-1-1 and its long illustrious history to a totally new architecture based on the use of IPs and SIP signaling.

Rick Jones was the operations director for NENA who brought NG9-1-1 concepts to the IIT VoIP lab through personal visits and conference support early in the NG9-1-1 development stage. Rick was part of the first NG9-1-1 teams and saw the benefits of technical engineering, programming, and data management that graduate students could bring to bear on the future of NG9-1-1.

The heart and mission of emergency services is indeed the public safety personnel who save lives every day around the world. Excellent communications capabilities enable the mission.

As we approached this mostly technical "how-to" book, it was with the goal of making the mission of deploying next-generation communications easier to accomplish. The inspirational people are too numerous to mention, so herein we acknowledge all of the thought leaders. We rely on the people who inspired the mission and educated us.

10.3.1 Telematics and Trials

A major threshold was reached when innovations such as telematics' direct connectivity to E9-1-1 systems and PSAPs were tested and found to be viable. NENA used another integrated E9-1-1 Rube Goldberg[8] adaptation to connect into that

6. The SIP Forum is a nonprofit organization devoted to advancing the adoption of the SIP, a signaling protocol for use in initiating, modifying, and terminating an interactive session among two or more users that involves multimedia elements such as voice, video, instant messaging, online games, and virtual reality
7. The IETF is an open standards organization that develops and promotes voluntary Internet standards, in particular the standards that comprise the Internet protocol suite (TCP/IP). It has no formal membership roster or membership requirements. All participants and managers are volunteers, though their work is usually funded by their employers or sponsors. The IETF started out as an activity supported by the federal government of the United States, but since 1993 it has operated as a standards-development function under the auspices of the Internet Society, an international membership-based nonprofit organization.
8. Reuben Garrett Lucius Goldberg, known best as Rube Goldberg, was an American cartoonist, sculptor, author, engineer, and inventor. Goldberg is best known for his popular cartoons depicting complicated gadgets performing simple tasks in indirect, convoluted ways. (*Source:* Wikipedia.)

platform. Refer to "Final Evaluation Report, Evaluation of MAYDAY/9-1-1 Field Operational Test" "prepared for the Federal Highway Traffic Safety Administration by Battelle in association with the Melcher Group, July 19, 2006. Its executive summary shares problems operating in an E9-1-1 world and work around solutions that were tested in Minnesota:

> The Minnesota Department of Transportation (MnDOT) conducted a federally funded Field Operational Test (FOT) to develop and test a MAYDAY/9-1-1 system. The system automatically routes 9-1-1 calls initiated by Telematics Service Providers (TSP) to an appropriate Public Safety Answering Point (PSAP) based on vehicle location information provided by the instrumented vehicles. This Field Operational Test (FOT) voice routing solution is called TSP Emergency Call Routing Service (TSPECRS). In addition, the Automatic Crash Notification (ACN) or Advanced Automatic Crash Notification (AACN) data from the TSP are shared in real-time with MnDOT in support of emergency response and management via the Conditioning Acquisition and Reporting System (CARS1).

Battelle, in association with the Melcher Group, was under contract to the Intelligent Transportation System (ITS) Joint Program Office (JPO) of the U.S. Department of Transportation (USDOT) to conduct an independent evaluation of the MAYDAY/9-1-1 FOT.

> The scope of this evaluation is to conduct an acceptance test on the voice routing system, TSPECRS; perform an analysis on ACN/AACN data routing, solicit feedback from the users of the voice and data routing systems, and identify potential deployment issues. This evaluation report documents the process, results, and issues encountered in this FOT, from which lessons learned can be drawn to provide guidance for the future deployment of this or a similar system. Implications for NG9-1-1: Third Party Technical Review In accordance with the evaluation plan, the FOT concept, technical design, and results of the field test were reviewed with Mr. Roger Hixson, NENA's Technical Issues Director. The interview specifically related to NENA's work in the area of Next Generation E9-1-1 systems, recognizing that the latter will ultimately generate or facilitate enhanced solutions for the objective of this FOT. Mr. Hixson observed that, while the design involved would require new TSPECRS-specific functionality in the Wireless network, a potentially challenging matter in light of evolving Wireless networks, the concept and solution being tested by the FOT represented a 'viable interim method' to automatically route emergency calls being generated by customers through telematics call centers to 'native Enhanced 9-1-1 systems in the United States.' Ultimately, Mr. Hixson indicated, IP based, NG 9-1-1 systems are likely to offer 'more effective and seamless solutions for ACN calls and basic data delivery, and subsequent PSAP access to supportive data from Telematics provider databases.

As recently as 2017, efforts were led by Senators Amy Klobuchar of Minnesota and Bill Nelson of Florida to fund what has been described as a new system to save lives. Refer to "Senators Aim to Boost Federal Investments in Next Generation 9-1-1 Technology," dated November 9, 2017:

> Existing federal grant programs would be expanded to help state and local governments deploy modern 911 systems that enable citizens to text for help and

> send audio, text or video files during emergencies under legislation introduced on Thursday." "As a former prosecutor and co-chair of the NextGen 911 Caucus in the Senate, I know how important it is for our first responders, law enforcement officers and Public Safety leaders to be able to communicate seamlessly during times of crises.

To date, needed funding to save lives on the highway as envisioned by Laurie Flaherty, and the DOT has not been forthcoming. Chapter 12 addresses traditional means of funding 9-1-1 and concepts for NG9-1-1.

10.3.2 Text to 9-1-1

Universal text to 9-1-1 using SIP-to-SIP signaling is achievable if the NG 9-1-1 core is in place and universal. NG 9-1-1 must ultimately provide seamless service that includes the accurate location information to render emergency services. The vision has yet to be achieved and has progressed as many 9-1-1 services have, requiring another Rube Goldberg bolt-on capability. How did we get to the current contraption?

Text to 9-1-1 was tested as part of NENA ICE 5, October 2012, at IIT as a sponsored ICE event. Hearing- and speaking-impaired individuals and leaders from NENA, universities, and companies were represented nationally and internationally to demonstrate that text to 9-1-1 would work. Sign-language interpreters were on hand for the participants, and the participants were able to witness calls originated from a specially adapted access solution on site in Wheaton, Illinois, to text 9-1-1 using a ISDN PRI trunking interface into the NGCS and through to the i3PSAP equipment. The "call taker" was able to bridge in language-support services, and the logging and recording companies were able to log the exchange of information into their systems.

On May 15, 2014, the FCC issued a mandate for interim solutions for text to 9-1-1 that was implemented early in a mostly nonstandard fashion because of urgent demands to support persons with disabilities; text to 9-1-1 was provided by the four largest wireless OSPs using any one of three NENA-approved options. A NENA fact sheet defines the options (https://transition.fcc.gov/pshs/911/Text911/NENA_FAQ_for_Interim_Text_911.pdf).

The three options included the following:

1. Text to TTY/TDD: The PSAP accepts the call from the traditional TTY/TDD equipment, but it displays in a unique way to allow the PSAP to recognize it as text to 9-1-1.
2. Web portal: The PSAP call taker opens the web page on a separate screen and keeps it open for shift duration. Vendors could incorporate that screen into their existing equipment.
3. Fully integrated NG9-1-1 with text becoming a new originating class of service: Read the NENA PDF for more detail and the full set of NENA-related documents when implementing the version chosen by the NG9-1-1 RFP winner.

The PSAPs are allowed to decide when they can afford to accept text to 9-1-1 and be prepared to advertise it within their communities. Since major wireless OSPs were required to offer the service, there was a need for TCCs to be built, and companies were ready to become TCCs; notably in Indiana Mark Grady and INdigital were already offering their own text to 9-1-1 solution supported by the Indiana PUC Wireless Commission. Similarly, Comtech TCS quickly began to offer TCC services. The FCC created a database and tracking program to allow PSAPs to post six months in advance of their planned readiness date for text to 9-1-1 service. In the interim, the wireless OSPs that received text to 9-1-1 queries integrated their SMS with the TCCs and tested and terminated to connected PSAPs. Until each PSAP is ready, the wireless provider is mandated to provide an intercept message to the effect that text to 9-1-1 is not operational in the area and that if emergency services are needed, it is necessary to immediately call or dial 9-1-1.

None of the access options strictly follow what was envisioned by NG9-1-1 where the caller's wireless handset can communicate through the network interactively with the PSAP call taker using SIP access protocols. Almost all wireless calls today arrive at either the E9-1-1 network or the NG9-1-1 network with SS7 signaling involved. A few major wireless OSPs do originate calls via SIP, but their aggregators predominantly use Level 3 as part of the transport process, and Level 3 converts SIP to SS7 for delivery.

SS7 is required for almost all E9-1-1 SRs. SS7 is adding costs, network delay, and more interfaces and equipment for NG9-1-1 whose native signaling is SIP. As with many issues, getting text to 9-1-1 to work, even under less than standard means, is the right solution until NG9-1-1 deployment becomes ubiquitous.

Those who plan to deploy NG9-1-1 and need to reintegrate and test the text to 9-1-1 solution should consult the important, if not mandatory, set of references from the website NENA.org, noting that the rules, regulations, and technology change over time. These references include the following:

- Test to 9-1-1 Public Education Plan Document, "NENA-REF-003.1-2015". Reference: "ASL/Captioned Videos About Text to 9-1-1" on "NENA You Tube Channel."
- "Interim SMS Text to 9-1-1 Information and Planning Guide," including Carrier Questionnaire, Request for Service Letter, Carrier Implementation and Testing Information, Carrier Points of Contact, for AT&T, Sprint, T-Mobile, Verizon Wireless. Availability May 15, 2014.
- FCC PSAP Text to 9-1-1 Readiness and Certification Registry," https://www.fcc.gov/general/psap-text-911-readiness-and-certification-form.

It is essential to add the specialized text to 9-1-1 test plans to your version of the Chapter 7 test documents.

10.3.3 Location Services

The vision of NG 9-1-1 was to have the exact location of the caller to deliver emergency services. This would include not only the lat/long (X/Y) coordinates but also the height, the Z coordinate. This vision is yet to be realized. Earlier chapters repeatedly reference location services and the pitfalls that this presents, especially for wireline and VoIP customers who are mobile. News events asking why a pizza can be delivered to a location, but the PSAP cannot locate a 9-1-1 caller, prompted the FCC and citizens to ask why 9-1-1 call takers could not consistently see where someone is located when they are pleading for help—in many cases when calling from wireless and VoIP handsets. Recall that wireline location services transition in a straightforward process to NG9-1-1. With the MPCs and VPCs and their remote databases and procedures, location accuracy for these customers is constantly improving and meeting milestones set forth by the FCC for QoS improvements. Enterprises provide the largest changes for locating callers and locations with large acreage and with multiple buildings on site such as barns and outbuildings. As with text to 9-1-1, NENA was faced with a decision to wait for full NG9-1-1 conversion, including large enterprises or agree to another interim measure. OTT[9] solutions, as described in Chapter 6, have been widely supported by nearly every PSAP and many of the owners of the onsite PBXs where the calls originate. RapidSOS is currently the leading provider with integrated solutions and a rapidly increasing partnership with OSPs, third-party providers, and customers. This definitely is a bypass of the NG9-1-1 network from enterprise through OTT equipment and databases to these large customers. As mentioned, in 2020, the FCC added to its location guidelines requiring GIS Z-level position information to be sent for 9-1-1. This is an initiative to find the floor the caller is on in multistory buildings. There have been many efforts tying databases of campuses and businesses to nearby Wi-Fi and/or Bluetooth devices to bring 9-1-1 calls through with additional information using the NG9-1-1 network. OTT solutions are competing with longer term standards for a chance to become a de facto standard. In any event, we cannot say how this will end, but consideration when converting enterprises to connect them to the NG9-1-1 network must include sample testing to comply with FCC location guidelines and provide adequate security and safety. As almost everyone knows, third-party providers support large enterprises when it comes to 9-1-1 location services.

10.3.4 Beyond the Call for 9-1-1

What NENA originally described was the vision of maybe a handful of people at most, for i3 or phase 3 of emergency services worldwide. The vision was for a worldwide solution no matter the digits dialed or the technology being used to request help. It can be expanded to fit the demands of 3-1-1 nonemergency services and the whole array of three-digit specialized access numbers. The core or heart of the network and systems is to be resilient, redundant, diverse, and reliable.

The caller and the call takers can be anywhere; physical walls need not bound a PSAP. The ability to transfer and serve will be ubiquitous, because all of the information from the caller keeps traveling with the call as long as the chain of next generation technology is not broken. The vision is to fully integrate access to the field of first responders or other service-oriented call takers , providing them with

9. Chapter 6 introduces the topic of OTT communications supporting emergency services. Specifically, OTT pertains to PSAP access of the location database of the major MLTS enterprise customers.

much more data including possibly biometrics[10] attained from medical devices and even smart phones and the telematics from vehicles where passengers are waiting to be triaged.

This book cannot address the complex decisions involved in approval of interfaces, or issues such as biometrics and telematics, which may have privacy issues associated with them. Because it is possible to send heart rate or other key medical information via a network to a first responder, it may open up privacy issues and other concerns. It is important to understand the full implications of accepting and storing and protecting information from being available to sources not authorized to receive it. In the case of telematics, the speed a car is traveling or the blood alcohol content of the person holding a biometrically capable steering wheel may be helpful in locating callers and prioritizing triage, it could be used in ways not envisioned, such as insurances companies and legal claims. So the future may have technical solutions that are challenging to implement in the real world. Thanks to NENA, APCO, and all the standards bodies that make these new services a technical reality. Those organizations are not the place to address the legal and privacy decisions associated with requirements.

10.3.5 Apps

There are many app developers that have paved the way for features and functions for emergency services in large part to help solve unsolved problems and in another way because there is a market for apps.

10.3.5.1 Access

Smart phones and watches have been popular new gadgets or tools for runners and people who want to remain hands-free and still stay connected. Biometric health data and related items are all stored in memory and uploaded to the Internet for tracking and personal development. A feature of some of these tools is the integration of access to emergency services. NENA has invited app developers to bring their products to an ICE event or lab to test for integration with developing standards. PSAPs often get calls from these applications and may or may not be expecting the calls or have the means to locate the caller. Users can accidentally trigger the emergency feature and cause a short storm of demand in PSAPS. There are going to be more not less of these apps developed, but their full positive impact will not be realized until fully end-to-end NG9-1-1 is operational.

The "Find my phone" feature found on both IOS and Android devices can be set up so people can locate their own phone or find others who they are connected to with this app or feature. Some see this as a safety feature and others as an invasion of privacy. Nevertheless, the feature could help a person when they are asked where they are located by the PSAP call taker who cannot see their location on the screen for any reason. If the phone user who needs help can take a snapshot of their location, they could text that photo to the 9-1-1 call taker using text to 9-1-1 and

10. Biometrics are body measurements and calculations related to human characteristics. Biometrics authentication is used in computer science as a form of identification and access control. It is also used to identify individuals in groups that are under surveillance. (*Source:* Wikipedia.)

possibly save their own life or the life of a someone else, a workaround that could save lives.

The number of PSAP applications grows exponentially with NENA referencing solutions daily on their briefings to members. Upcoming ICE events will test these kinds of apps with willing participants.

10.3.5.2 NOC as an App

NENA has been developing a standard for monitoring the NG9-1-1 network. (See Chapter 9.)In CSI, Assure911 developed and deployed an NG9-1-1 app for Androids and IOS to provide integrated surveillance and alerting. A physical NOC could use this as an adjunct capability to their physical NOC so NOC managers on the go can be alerted to events that could leave a 9-1-1 service in simplex conditions or show that there is a problem in the NG9-1-1 end-to-end network. PSAPs and managers could benefit from a hand-held app to see their calls are being diverted to an alternate PSAP and help them with staffing decisions in the event of loads or overloads. This app was integrated during selected ICE events and presented at various forums. An article was published in *IEEE Communications Magazine,* January 2017, Volume 55, No. 1; the issue was dedicated to emerging issues such as NG9-1-1.

IP-enabled security cameras such as Ring and others are becoming more widely deployed by homeowners and business. These cameras can be integrated into public safety solutions. Privacy concerns abound around this issue as to who has access to the devices in real time or access to the recordings. Ring has an option to share recordings with police for specified incidents. Increasingly in the United States and especially worldwide, surveillance cameras and other monitors (such as gunshot detectors) are being deployed. Access to these devices by public safety to respond to emergencies will be achieved once the NG 9-1-1 core is deployed and accessible by the first responders.

10.4 Forums and Standards Bodies

Forums and standards bodies are responsible for bringing forth white papers and proposals to move research and development forward. The SIP Forum sponsors many requirements for SIP protocol, the backbone of the NG9-1-1 network. NENA addresses the core of the NG9-1-1 system, while APCO tends to address the PSAP and egress through FirstNet to ensure that first responders' requirements are addressed. Access standards bodies abound from 3GPP and more for the originating technologies and services. Other organizations include the IETF. Also, with an active portfolio of nearly 1,300 standards and projects under development, the Institute of Electrical and Electronics Engineers (IEEE) is a leading developer of industry standards in a broad range of technologies that drive the functionality, capabilities, and interoperability of products and services, transforming how people live, work, and communicate.

10.5 Roles of Universities

University leadership has been vital to set the vision for and bring the future of emergency services into reality. In the past what was brainstormed and brought to fruition by Bell Laboratories, Bellcore, and Telcordia, and their successor organizations is now being accomplished through universities working in partnership with government agencies and industry leaders and multiple standards bodies to accomplish the goals

Universities and their laboratories are providing a vendor-neutral environment in which architectures and applications can be tested to assure that they meet the many 9-1-1–standards based requirements. Key contributors include the following:

- Professor Henning Schulzrinne of Columbia University, New York City, NY;
- Professor Walt Magnussen of the Texas A&M University, Austin Texas;
- Adjunct Professor Carol Davids of the Illinois Institute of Technology.

Columbia University leaders and students developed many of the first prototypes of NG9-1-1 systems based on draft standards. Schulzrinne went on to become FCC CTO for a time and added to the body of NPRM documents, including those to improve network reliability and diversity after major storms and outages disrupted 9-1-1 service. In one rulemaking, the 9-1-1 SSPs are now required to notify PSAPs if their service is interrupted so they can move to alternate facilities and or activate plans to have calls answered by other agencies while their systems, facilities, or service are impaired. Schulzrinne has written books on the topic, advancing the reality of NG9-1-1. The FCC is responsible for: (1) wireless 9-1-1 requirements; (2) added VoIP 9-1-1 requirements; (3) managing the Network Outage Reporting System (NORS) and database; 4) record-keeping and outage analyses; and (5) the database for text to 9-1-1 PSAP readiness, among many other projects. Schulzrinne remains a stalwart advocate for NG9-1-1 and speaks regularly at conferences and events nationally and internationally.

Walt Magnussen hosted the POC Testing at Texas A&M for NG9-1-1 sponsored by the DOT, led by Laurie Flaherty. Magnussen is immersed in DHS projects for NG9-1-1 and actively supports the Commission on State Emergency Communications (CSEC), the Texas NG9-1-1 project, among other systems and projects. His team has a deep knowledge of the NG9-1-1 GIS and database capabilities. He is a regular speaker nationally and internationally for NG9-1-1 and related matters. He is also supportive of FirstNet projects and has published articles for IEEE and others to assist in promulgating NG9-1-1 and FirstNet deployment. His team hosted early NENA ICE testbed events as noted in Chapter 7.

Carol Davids built and manages the IIT RTCL labs in both Wheaton and Chicago, Illinois. Her first NG9-1-1 efforts came about as part of graduate student projects, requiring the students to build and manage private and public LoST servers. In 2009, Davids assisted Clearwave and the Counties of Southern Illinois (CSI) in their BTOP grant application for a "shovel-ready project" to be among the first in the nation to deploy a NG9-1-1 network. In 2010, using students from Columbia, Texas A&M, and IIT, graduate students did a live demonstration at an IIT RTCL conference to demonstrate that standards for NG9-1-1 are real. Calls were made from the auditorium using lab resources connected over the Internet to equipment in New York City, Austin, and Wheaton to call and text 9-1-1, and locate the caller using Google map services. The Wireshark traps and traces were shown as well as the PSAP call answering. The graduate students working independently then together created the project in about two months. Davids has been a member of the National Emergency Address Database (NEAD)[11] team researching the means to improve indoor accuracy for NG9-1-1 in multistory and cross-campus locations. Students add to the research work each semester and demonstrate their techniques, changing vendor-provided cell phone software, adapting and installing WiFi and Bluetooth devices, and creating location databases, working with wireless industry partners. The IIT RTCL accepts vendor, industry, bilateral, and integrated test projects to improve student knowledge and provide another place for small and large vendors to test their product resilience and ability to achieve seamless integration. The IIT RTCL hosts newer NENA ICE events instantiated by Texas A&M and continues to be a premiere destination for these events as noted in Chapter 7. Davids is an advisor/member for IEEE and a member/contributor of IETC, Cable Labs, and notable forums. IIT was instrumental in specific testing for NG9-1-1 for CSI. Davids and her graduate students devised testing protocols to attack the NG9-1-1 network before it was operational to find any flaws.

As mentioned in Chapter 7, laboratory load and overload testing before live traffic is added to any ESInet, NGCS, and PSAPs is essential to prove parameters and settings are adequate for the SBC and other capacity limits are in place. The last thing any network needs is a system that cannot perform in a storm or disaster when it is most needed. Cybersecurity is tested by simulating attacks from within the network and from the outside to verify the alerts including that PSAPs, NOCs, and security teams are notified in real time and can adapt and adjust to the situation. Without third-party neutral confidential testing, this was not possible.

The IIT lab has created virtual PSAPs and advanced the concepts of NG9-1-1 to a point where virtual PSAPs could be a part of the future vision. The lab has deep expertise in testing and signaling protocols for the telephony circuit-switched world and for SIP, RTP, and next-generation technologies. When teaching students, Davids bridges the gap from the PSTN to the IP networks and the future. Davids teaches masters degree programs, speaks nationally and internationally, and is well-published

10.6 Egress—NG9-1-1 and FirstNet

As NG9-1-1 moves forward on its own trajectory in the United States and internationally, APCO, a sister public safety organization to NENA, has been consistently moving the concept of FirstNet forward since the attacks on September 11, 2001. The differences are stark in some ways because a single event and the subsequent 911 report created a demand for integrated first responder communications, and Congress funded the project. An RFP was released (and subsequently won by

11. NEAD, LLC, was established by CTIA-The Wireless Association (representing NENA, APCO, and 4x Tier 1 Carriers), which designs/implements the NEAD. The NEAD stores the physical addresses of WiFi and Bluetooth beacons to determine the "dispatchable" location of the beacon itself.

AT&T) after extensive requirements were developed and because of funding and legislative mandates to implement the project, the work has begun in earnest. NG9-1-1 did not receive any similar mandate or funding, nor has a single organization been assigned to manage and track deployment in the United States. The question is, how will these two projects come together and what is the relationship. Articles published in 2015 offer some insight. Refer to Responder News: An Overview of FirstNet and NG9-1-, July 24, 2015:

> A report issued by the National Public Safety Telecommunications Council (NPSTC) provides an overview of two public safety communications networks now in development: the nationwide public safety broadband network being advanced by the First Responder Network Authority (FirstNet), and Next Generation 9-1-1 (NG9-1-1).[12]

Since FirstNet and NG9-1-1 share the goal of improving public safety communications through IP-based architecture, they are natural partners. While NG9-1-1 will enable PSAPs to receive more detailed information than ever before, the FirstNet network will enable those PSAPs to share the information with responders in the field and give responders the capability to share critical information among themselves. Many public safety practitioners anticipate these networks will dramatically enhance the speed, effectiveness and safety of response operations."

"FirstNet and Next Generation 9-1-1: High-Level Overview of Systems and Functionality," can be viewed online.

10.7 Summary

The good news is that NG 9-1-1 has been not only proven to work in labs but has also been deployed in the real world with demonstrably superior results in performance to the PSTN based E9-1-1. The sad news is that the vision is a long way from being met. The PSTN still bottlenecks the technology to achieve SIP-to-SIP connectivity, unleashing the planned and unimagined capabilities that could be realized. NG 9 -1-1 needs support with political leadership on both the national and state level. Funding has been sorely lacking, and to achieve the original vision of NG 9-1-1, it's going to take real political courage to allocate the type of money required to fully deploy NG 9 -1-1. There is so much more we could say about emerging technologies. Technology is leaping forward. The call takers know what they want and need. There is no lack of dedication, but the more workaround solutions that are deployed, the harder it is to achieve seamless standard processes that are essential to the public safety mission. Public safety agencies need to focus on public safety *not* technology. The call takers need a seamless integrated environment for work and a simplified way to reach the right first responders. We are all derelict in our mission to help them achieve their goals. The information in Chapters 1–9 will be eclipsed in a few years with even more amazing solutions. Public safety deserves a chance to catch up and get ahead of the curve not only in the United States but worldwide.

12. https://www.dhs.gov/science-and-technology/news/2015/07/24/responder-news-overview-firstnet-and-ng9-1-1

CHAPTER 11

Legal and Regulatory

11.1 Lawyers

Lawyer up! While the telecommunications business has been deregulated over the years, this is not the case for 9-1-1, which is regulated by both the federal government and most state governments. In addition to the regulation, there is a great deal of liability and responsibility that must be addressed. This chapter describes the regulatory challenges encountered in several areas of the United States and provides analyses to plan a successful course of action. The NG9-1-1 project teams must know the legal and regulatory rules and requirements that support and inform their decision making.

A quick discussion of lawyers. They all went to law school and passed the state bar exam. Among them, however, those factors are where the similarity ends. Because of the expanse and complexity of the legal system in the United States and internationally, lawyers generally specialize in certain areas of the law. Some lawyers can fill multiple roles necessary to implement a NG 9-1-1 system. Regulatory lawyers are critical in dealing with state and federal regulators. Delays can be encountered if there are protracted regulatory proceedings. Seeking approval of a NG 9-1-1 project can involve multiple filings with the regulators, and there may be opposition to the plan from a displaced 9-1-1 SSP or other parties. Regulatory skills can be key in negotiating OSP and 9-1-1 SSP agreements. Contract law is another skill needed for public safety agency and vendor contracts. Lawyers employed by the government who work in public safety tend to be experts in criminal law, which has limited application in installing a NG 9-1-1- system. There are times the public safety agency will want to use in-house counsel. Some public safety agencies have outside counsel and that may provide access to legal experts in required areas of the law. All parties should be aware of the areas of expertise and kinds of legal requirements that need to be addressed so these expensive resources can be used with maximum efficiency.

One other note on working with lawyers. Brilliant as they may be, they are experts in legal matters, not the drivers of the project. Lawyers should be involved in the initial development stages of the project with "here's what we want to do" instead of asking "what should we do?" Lawyers are legal experts not visionaries.

The "needs of the business" should define the direction and scope of the project. Lawyers help determine how to get there and what to avoid.

11.2 Federal and State Jurisdiction

The FCC states on its web page[1] "911 service is a vital part of our nation's emergency response and disaster preparedness system. In October 1999, the Wireless Communications and Public Safety Act of 1999 (911 Act) took effect with the purpose of improving public safety by encouraging and facilitating the prompt deployment of a nationwide, seamless communications infrastructure for emergency services. One provision of the 911 Act directs the FCC to make 911 the universal emergency number for all telephone services." The 911 Act extends the supremacy clause of the U.S. Constitution, giving the FCC ultimate authority for 9-1-1. Historically, in the telecommunications industry, the FCC has jurisdiction over interstate traffic while the individual states maintained jurisdiction for intrastate traffic. Without a lengthy discussion of why, the FCC delegates the regulation of wireline 9-1-1 to the states while retaining complete jurisdiction over wireless and VoIP services. The states usually regulate the telecommunications carriers, including the 9-1-1 SSP as well as the public safety agencies and their governing boards.

The FCC has required wireless and VoIP providers to terminate their 9-1-1 traffic at a SR or 9-1-1 tandem. This split jurisdiction should not cause implementation issues for NG 9-1-1 unless the definition of SR or 9-1-1 tandem comes into question. The terms SR and 9-1-1 tandem are E9-1-1 era language. To avoid any issue with this regulatory rule, the references to the ESRP and to a larger extent the NCGS can be cobranded as the IP SR to comply with the FCC rules for wireless and VoIP.

11.2.1 Statutes Versus Regulatory Rules

As every high school graduate should remember from civics class, legislation, known as statutes or laws, are created, debated, and passed by legislatures and signed into law by the executive. Passing legislation can be a lengthy and tedious task. Because of the difficulty representative bodies face in dealing with and reacting to all the issues that confront the country or state, especially in the area of technology but not limited to it, the legislatures both nationally and in every state have created agencies to oversee and regulate industries of common public interest and importance. The responsibility to staff and manage these agencies has been delegated to the executive. An important power given to the agencies is to promulgate rules, known as regulatory rules, which have the power of statutes and prevail unless there are statutes that conflict with the regulatory rule. In case of conflict, the statutes always prevail. The state organizations dealing with 9-1-1 and/or emergency services are often called public service commissions (PSCs) or public utility commissions (PUCs), and in some places the state police agencies carry authority over 9-1-1. In some cases, the county and/or state may manage the 9-1-1 public service agencies.

1. https://www.fcc.gov/general/9-1-1-and-e9-1-1-services.

The point is that wherever the reader is located, determine if there are oversight or governance issues for deployment of emergency services, learn the rules, and identify the legal and regulatory advocates.

11.3 Tariffs

Most people, even those in the telecommunications business, do not fully understand tariffs. The common belief is that tariffs are engraved in stone and cannot be changed. This is a total myth, and any excuse for not being able to do something because of "the tariff" is a smoke screen, a stalling tactic, or out-and-out obstruction. In the telecommunications business, in almost all states, companies that want to offer services to the public that fall under the jurisdiction of a regulatory body are required to file a tariff. This document defines the services belonging to the public. By public, this typically means that any customer, even if they are unsolicited, can purchase products and services out of a tariff if they qualify for the terms and conditions outlined in the tariff. An example would be that only governmental entities (local, state, and federal) can purchase 9-1-1 services offered by a 91-1 SSP's tariff. In states that regulate 9-1-1 SSPs, some may require the 9-1-1 SSP to file a tariff for the services they offer. OSPs are required to publish their tariffs on the company's website.

The tariff by, definition, includes pricing for the tariff-defined services. These prices are available for all qualified customers who purchase from the tariff. The term of service is typically a month-to-month term and can be terminated at any time. Because these services are offered to practically anyone on a month-to-month basis, the tariff price is the price ceiling for the services. Practically all tariffs have language that allows contracts to be negotiated for the services at reduced prices from those listed in the tariff and always have terms of one year or more. Typically, 9-1-1 services contract terms extend from 36 months to 10 years.

Tariffs are essentially similar to legislation or as a set of semiregulatory rules. One thing all tariffs contain is language about the limitation of liability, indemnifying the company issuing the tariff against lawsuits for service failures, including loss of life. Refer to this example

> Company, its officers, directors, agents, servants, and employees shall not be liable to Customer or any Person for, and shall be fully indemnified and held harmless by, Customer against any claim or loss, expense or damage, for defamation, libel, slander, invasion, infringement of copyright or patent, unauthorized use of any trademark, trade name, service mark, or proprietary or creative right, or any other injury to any Person, property, or entity arising out of the material, data, or information transmitted.

Company, its officers, directors, agents, servants, and employees shall not be liable to any Person for failure to provide service.

No agent or employee of any other carrier or company shall be deemed to be an agent or employee of Company.

Company, its officers, directors, agents, servants, and employees shall not be liable to Customers or any other Persons for any incidental, indirect, special or consequential damages of any kind, including personal injury, loss of use, loss of business, or loss of profit.

The agreement between Company and the Customer is exclusive; there are no third-party beneficiaries."

Tariffs benefit the telecommunication companies that write them. Changing a tariff is not a complicated issue. Typically, the process to change a tariff is submitting the pages that are being modified with updated version numbers and the date of modification. Depending on the state and type of service the tariff changes may automatically take effect or in some cases the changes need to be reviewed and approved by the regulatory bodies. Plan on changes to 9-1-1 tariffs to be reviewed by the regulatory agency(s).

11.4 Requests for Information (RFIs), RFPs, and Contracts

The response to the RFI or RFP is not critical from a legal standpoint unless there is misrepresentation of facts. It is assumed that all parties are honest and are acting in the interest of all stakeholders. All team members need to be aware of the realities of the world we live in and adopt the old saying of "trust but verify."

The 9-1-1 SSP that replies to the RFP should make its responses clear as to what requirements it can conform to 100% and where there are exceptions on a sliding or absolute scale.

11.4.1 RFP as Part of the Contract

Part of the RFP process is the bidder responding to the RFP with specific responses and perhaps a note on each requirement that their company can meet or not meet. Of course, these responses are used to evaluate and pick the winning bid. Provide specific instructions and requirements in the RFP to avoid misunderstandings. The RFP should be completely incorporated in the final contract with the public safety agency and the prospective 9-1-1 SSP. This saves time to restate all the RFP requirements and what each party is responsible for and the intention or ability to address each requirement. Chapter 7 advises that test cases be built based on the RFP requirements, making the RFP response even more critical should a dispute arise. The RFP can be consulted as a prime source and if it is incorporated in the legal agreements; a formal process needs to occur to alter the project requirements.

11.4.2 Vendor Agreements

The RFP should also be included in vendor contracts similar to the public safety agency contract. The RFP should outline the requirements for each vendor and component or service that vendors provide. The RFP is the guide to specific vendor contracts. It is not advised to use a boilerplate for vendor contracts. Because of the specific nature of each component(s) the vendor provides, the contract should only contain requirements and SLAs for that vendor. The interpretability of the component(s) should be detailed in the contract, and the RFP reinforces the functions and responsibility for the vendor. Recall that SLAs as referenced throughout the document only carry weight, including penalties, if they are stated clearly in the contractual agreements. It is best to include any performance penalties upfront in the RFPs if they are not explicitly stated elsewhere and refer to them explicitly when negotiating the vendor contract for both products and services. That includes all aspects of the 9-1-1 service end-to-end. If the FCC and/or the state include reporting requirements and specify timing for those reports, be clear in any contracts who is making the reports and what the vendor must do to help, if requested, in a time of network stress or time for reporting afterward. Include the requirement to participate and sign off on MOPs and assist with required RCAs after an event occurs in a timely manner in support of regulatory rules and obligations. If penalties for non-performance include financial obligations, rebates, or credits, be specific and exacting on how these penalties are to be measured, over a month or an event or both.

11.4.3 SLAs

SLAs are performance expectations based on metrics and standards and are part of the contracts or agreements for physical network, software, and databases. The main purpose of the SLA is to maintain a level of expected performance.

SLAs are specific deliverables included in the contract that are measurable and achievable and have actionable remedies if the metrics exceed the parameters of the contract containing the SLA.

11.4.3.1 SLA Samples

Important technical specifications include the following:

- Minimum bandwidth;
- 1.544 Mbps on ingress from any interconnected carrier and egress to any PSAP;
- .5 Mbps per position to any PSAP (number of positions can grow—refer to design planning guidance in the NENA documents).

In addition, each data center shall be engineered to carry no more than 40% of the total demand of the NG9-1-1 network. This assumes that if there is a failure of one or more of the data center core FEs or of the total data center, the load to be carried by the remaining data center is at 80% of the normal demand. If the major event is caused by an extraordinary event, a single focused overload or a general demand overload, or a planned downtime of a part of the network, the remaining 20% could carry the full load until the network is restored. The ESInet needs to be designed with these limits in mind.

In terms of QoS (with alerts to monitoring if exceeded), the following parameters need to be considered.

- Packet loss: Budget for ESInet data center core equipment to the PSAP to be 2.5% with no ESInet over-subscription and 1% for ESInet transport.
- Jitter: There should be less than 20 mS in the end point jitter buffers;
- Latency: End-to-end - </= 150 mS (ITU-TG.114) and ESInet portion </= 35 mS packet delay;

QoS measurements will be monitored in the data center NOCs and exceptions sent to the 9-1-1 SSP monitoring system for alerting, ticketing, and reporting. Other information about the overall network will be discussed with the NOC manager and 9-1-1 SSP technical lead to refine the monitoring system details.

In terms of diversity and redundancy, the following should be considered.

- Generally, no SPOF should be part of the design.
- All PSAPs will have diverse fiber-terminating equipment.
- All PSAPs will be connected to the main fiber ring (not fiber string).
- All PSAPs will be supported with power that is documented and has NOC surveillance in the field and in the data centers.
- Power will be supplied as commercial, UPS, generator, or battery (as needed) in each data center.
- Power in the field will have, where possible, commercial, generator and battery (as needed).
- Power diagrams will be provided.
- Power will be monitored in the NOCs and results shared with the 9-1-1 SSP through the interface to the monitoring system for alerting ticketing and reporting.

The following are exceptions to the above specifications for diversity and redundancy:

- PSAPs with three positions or less with the ability of other PSAPs to accept the load in near real time via the ESInet ring and process them to designated PSAPs may be served by a fiber string or copper T1.
- PSAPs that share fiber-termination equipment may be acceptable if the PSAPs agree to the conditions. This would be for PSAPs that are small, and where the next routed PSAPs could handle the load for an indefinite period of time while restoration is taking place. (Note that fiber restoration is not a quick process.) These situations will be in writing. Cost estimates must be provided for alternatives, and all avenues must be exhausted.
- It is recognized that in the case of large PSAPs, it is ideal to have dual entrances to the PSAP. The entrance to the PSAP must be acknowledged as a SPOF, and that heightens the requirements for ESInet diversity.
- In addition, OSPs are requested to have two diverse paths to the data centers. There is no firm number of OSPs. As soon as OSP meetings are established the number can be refined. The following must also be considered when planning for diversity and redundancy.
- OSPs may refuse to diversely connect to the ESInet. The PUC/PSC staff will determine whether that is acceptable.
- In any event, even if one route is planned, and in the event of a failure of one data center, a component in the collocation space of the data center, or of the fiber at any point, that OSP must fail over to the backup equipment, transport route, and/or data center to reach any/all PSAPs.

- In response to other concerns about OSPs and the need to gain access to terminate their equipment, the 9-1-1 SSP will schedule each order and the delivery of the equipment and full information about the OSP's schedule with the installation coordinator in each data center in advance of the installation. Access for testing, maintenance, and cutover will be arranged through the public safety agency. The time windows for these activities will be known in general about 60 days ahead of the date, refined within the week of the delivery, and coordinated to the block of time and name(s) of the personnel doing the work.
- Installation expectations from the 9-1-1 include fire-suppression equipment and markings in the data centers, ESInet hubs, POIs, and PSAPs
- The 9-1-1 SSP should have secure and authorized access to the data centers and PSAPs both physically and logically pursuant to rules and regulations.
- OSPs have metrics for performance for NG9-1-1. Those items are their responsibility for reporting and managing for the FCC and PUC/PSC.
- SLAs will be established separately between each OSP and the new 911 SSP in accordance with PUC/PSC Guidelines and rules.

Design criteria and meetings expectations that should be included in SLAs are described as follows.

- Semi-annual design review and planning meetings shall take place between the data center managers and the CTO of 9-1-1 SSP.
- Actual measured performance for bandwidth and connectivity shall be discussed.
- Growth planning will be addressed in the event other entities join the NG 9-1-1 system.
- When the capacity of any leg of the network or component exceeds 60%, immediate action will be taken to expand bandwidth and or to augment the components necessary.
- NOC-to-NOC performance and notification will be reviewed along with major trends in outages.
- After the initial cutover, performance of the network will be reviewed to set the stage for any corrective activity in case the old network was being blocked or the new network is not performing as planned.
- Before the network is turned up, major overloads will be generated through the network looking for any problems before live traffic enters the ESInet or any FE. The data centers and the ESInet bandwidth providers will be notified of all testing and asked to help select the time and methodology for such DDoS testing.
- The overall network and OSP test plans will be shared with the data centers and bandwidth providers after drafts are prepared.
- Information suitable to support the PUC/PSC and other approval process(es) should be solicited in an ongoing manner until approval is gained.

11.5 Logging and Recording

The legal and regulatory requirements vary by state, but all public safety agencies record calls and need to retrieve calls. Vendors that provide logging and recording systems offer MIS systems to analyze and retrieve information about system events. Recall that the ESRP often provides MIS capability and the ESRP data is often incorporated into the logging and recording system that resides in the core. The systems that provide the NG9-1-1 service must interoperate according to the regulatory requirements with the storage systems in place.

11.6 OSP Agreements Versus Interconnection Agreements

Interconnection agreements are a product of the Telecom Act of 1996. They are what the title suggests, the interconnection of two networks to exchange traffic. Regulatory rules and statutes cover interconnection agreements; this is how competition was mandated for the major telcos and the telecommunications business. Interconnection agreements have a process for approval. Often the state regulatory agency must approve interconnection agreements, and part of the process is that if agreement cannot be reached, the parties must proceed to formal arbitration. To arbitrate the dispute between the two parties, the regulator becomes the arbitrator.

The interconnection agreement process is burdensome and is often used as a roadblock to obstruct progress by the embedded 9-1-1 SSP, which has little interest in converting to NG9-1-1 because it is losing the business or simply does not want to do the work. To stay out of the" briar-patch" of interconnection agreements, the authors suggest using "carrier agreements" to establish a contractual obligation of receiving traffic from the embedded 9-1-1 SSP. The 9-1-1 SSP will not be "exchanging" traffic with other companies; instead, 9-1-1 is a terminating service and does not require an interconnection agreement, which can delay a project. If the 9-1-1 SSP has existing interconnection agreements, they will cover the delivery of OSPs 9-1-1 traffic.

No two areas are exactly alike. Every OSP, large or small, needs a carrier agreement. Money has often been paid to RLECs or ILECs when negotiating connection agreements. In the case of NG9-1-1, sometimes the PUC/PSC dictates payment for services. In other cases, this can become a protracted agreement process. Smaller companies often want to be reimbursed for changing their facilities or signaling to terminate or augmenting trunks to one or in fact as desired two different points along an ESInet. Where facilities or access aggregators are involved, one agreement may cover a plethora of smaller VoIP providers or even CLECs. Refer to Chapter 12.

11.6.1 Transport Providers for Signaling Services

The RFP may or may not specify that the SSP requires multiple call signaling vendors to be involved as a subcontractor to the 9-1-1 SSP. Other vendors such as the map, workstation provider, logging, and recording vendors are visible to the public safety agency. Vendors such as transport/signaling providers need to be committed to the project in similar contractual terms as the other vendors. Including the RFP in the contract may or may not be appropriate depending on the call-handling details specified in the RFP.

SS7 contracts are almost certainly required for access. Often third-party SS7 providers are involved and need education on NG9-1-1. Data such as PCs for each access location that connects to the NG9-1-1 data centers, LNGs for signaling and trunking. Size of the links and full end-to-end diversity is a matter of negotiation as well as NOC access to the information to participate in rapid problem-solving. Note that, typically, the NG9-1-1 equipment in use today does not support virtual or Sigtran[2] SS7 links. SLAs for SS7 providers are different than the typical OSPs. The B or D links may be in place between points or need to be added or augmented for the 9-1-1 traffic. Certainly, penalties for service disruption of SS7 is a serious matter as well as 24-by-7 secure access to resolve problems. The CLLI and SS7 PCs and addresses for facilities to terminate are essential to any contract.

11.6.2 Specific ILEC Issues

ILECs send traffic from split exchanges to the primary 911 SSP per commercial agreements. Split exchanges agreements have a mutual responsibility clause. Typically, the 911 SSP that serves the most 9-1-1 lines from that split exchange NPA NXX is responsible to accept all the calls and return or transfer the calls that are outside that 911 SSP's PSAP boundaries (see Chapter 5). There are noted exceptions. In the NG9-1-1 network the NGCS that performs the return routing function is the LSRG. It is essential that the information to be divulged and tested be negotiated and that adequate trunking interfaces are provided to the neighboring ESInets or legacy SRs for call handling. Diagrams and details are essential. This is both a legal issue and technical issue. Eventually the systems will all be seamless when full ESInet to ESInet and GIS database services are in place. This is a costly and time-consuming detail. Parties may choose to opt in or out instead of dealing with split exchanges. This may require changes to interagency agreements.

11.7 Intergovernmental Agreements

Before beginning testing for an NG9-1-1 project, public safety agencies need to complete intergovernmental agreements (IGAs) for call handling and system procedures with emergency service providers. IGAs are contracts between two or more governmental entities. In the case of NG 9-1-1 these agreements between public safety agencies outline in detail how calls are handled between the agencies. The issues may include call overflow and backup agreements; dispatch of police, fire, and medical resources in certain areas of the footprint; call transfers; revenue; and expense distribution.

The agreements form the basis for linkage, cooperation, and communication between both intercounty and intracounty agencies. Informational meetings should

2. The IETF has defined the SIGTRAN protocol suite that implements levels 2, 3, and 4 protocols compatible with SS7. Sometimes also called *pseudo SS7*, it is layered on the SCTP transport mechanism. Some NGCS and LSGR and LNG vendors do not allow for a SIGTRAN interface for lack of equivalent reliability for critical A link carrying facilities interface.

be held with the stakeholders to develop a spirit of cooperation and the problem solving required for success.

When PSAPs back each other up singly or within an ESInet, typically each county needs separate agreements with the other. The same is true of neighboring agencies. Some of these agreements are in place, yet not always updated as to opted in and out arrangements, radio frequencies, administrative lines, etc. Review the data in Chapters 2 and 4 on the transfer protocols and the explicit diagrams developed for the project. They provide the details required to complete legal documents in a timely manner. Our experience has been very low-accuracy in the case of split exchange agreements. In some cases, there are payments required if one agency permanently backs up or serves as the PSAP for another agency. If a short-term outage or overflow occurs, almost no costs are exacted. If a PSAP is down a long time due to storm, terrorism, or damage, or even health reasons such as the 2020 Covid19 virus, the agency supporting the primary agency may demand reimbursement. Mutual aid agreements are of value before these situations occur. Work at home agreements and guidelines may be needed with the use of remote portable workstations.

11.7.1 Regional NG 9-1-1 IGAs

Forming a regional or statewide NG 9-1-1 project is entirely dependent on the legislation and regulatory framework in each state. In some cases, the state is able to direct the formation, configuration, fund and oversee the overall project. In other states the public safety agencies are more independent and able to form their own projects. Some regional efforts have the various public safety agencies form a tax exempt "not for profit" 501.(c).(3)[3] organization to operate. Others use a form of memorandum of understanding (MOU) that is nonbinding but that includes the requirements and responsibilities of each member organization. States where the public safety agencies operate more independently, absent a statewide plan, may require regulatory approval for 9-1-1 changes. Some states recognize regional organizations, while states like Illinois do not recognize regional organizations and require the individual public safety agencies to seek approval separately.

Regardless of the manner the regional NG9-1-1 organization is formed, the following issues must be addressed:

- Membership qualifications;
- Cost of entry;
- Ongoing cost sharing;
- Governance;
- Term of commitment .

Where IGAs are utilized in regional projects where two or more public safety agencies pool resources and issue a single RFP to build a regional NG 9-1-1 system to share resources and to connect the agencies for improved service, certain issues must be addressed and memorialized in a legally binding agreement. The regional

3. https://www.irs.gov/charities-and-nonprofits.

or state project IGAs address the issues listed above and also need to address in detail system governance and responsibilities.

Items frequently covered in IGAs (aside from standard contract language) are listed as follows.

- Formation and responsibility;
- Purpose (i.e., dispatch services - dispatch center, and default PSAP);
- Financials for 9-1-1 and billing budget (RFP costs and procurement costs);
- Financial oversight;
- Governance structure (voting rights of members);
- Governance committees (policy and operations including common processes);
- Technical sharing and costs (handoff, radio and administrative communications, PBX, cables, cabinets, data center costs, and ESInet);
- Meetings (schedule and frequency);
- Project leadership.

11.8 Summary

Each installation of a NG 9-1-1 system is unique, and the project leaders must review the prevailing legislation and regulatory rules. The 9-1-1 landscape is rapidly changing in the 21st century, and legislators and regulators are increasingly becoming aware of the need to upgrade the public safety infrastructure. Legislation to promote NG 9-1-1 continues to advance, and it is hoped that the trend will favor adequate oversight without burdensome regulation and restrictive legislation. The research necessary to review the legal and regulatory environment does not require a lawyer. Most people should be able to review the statutes and rules with a simple Internet search. Lawyers understand the rules of practice before the regulatory bodies, but most of the regulations and legislation are in plain language.

Some individuals often overlook or fail to deem important the legal matters of a NG 9-1-1 project. The phrase "no problem" should be banned from the project. Any legal issue that arises can delay or cancel the best designed and technically sophisticated system. Lawyer up!

CHAPTER 12

Financial

12.1 NG9-1-1 Myths

"No Bucks, no Buck Rogers." This quote is from the 1983 movie "The Right Stuff," which depicts the early years of the American space program; its point is that if there is no money for the space program, there can be no adventurers like Buck Rogers, a heroic character from 1920s science fiction.

Funding is undoubtedly the biggest obstacle to NG9-1-1 adoption. The cold reality is that NG9-1-1 deployment is expensive and time-consuming. In the deployment of the first rural NG9-1-1 system at CSI, one of the biggest challenges came from myths about NG9-1-1, which persist. A few of the myths are described as follows.

- NG9-1-1 is cheaper than E9-1-1. The thinking here is that "the phone companies" are ripping us off for E9-1-1, which is comprised only of computer parts off-the-shelf. The belief is that if each PSAP buys its own equipment for the back office today, then sharing the costs of the core equipment across many PSAPs is less expensive than E9-1-1 if they must upgrade anyway.
- It will not take very long to go from E9-1-1 to NG91-1. It is just computer parts assembled like we already have.
- Getting OSP traffic from the telco SR and having IP-enabled PSAPs is a NG9-1-1 system as if the OSPs are willing to reconnect to new locations and reroute their traffic away from legacy SRs for a faster more reliable direct connection.
- There will be an abundance of federal money and surcharge dollars to support NG9-1-1.

12.1.1 Dispelling Myths

NG9-1-1 myths complicate matters from the preparation to release of a RFI and RFP to contract negotiations and cutover. Because of the advantages of NG9-1-1, the goal is to get to implementation as soon as possible and at roughly the "same cost" as the move to E9-1-1 was in the first place. Vendors promote the idea that this can be done quickly once the contract is signed, but the costs and timelines are not the same as E9-1-1. While NG9-1-1 is based on existing IP technology that business and government have installed for years using off-the-shelf hardware, the NG9-1-1 design calls for redundancy, physical diversity, core services, and ESInet, all interfacing with new or embedded Maps, CAD software, and recording systems.

The redundancy of the design is essential to achieve the promise of reliability, as is physical diversity. This requires multiple (at least two) locations for the NGCS or core and the requisite facilities to provide the redundancy. Some FEs such as the BCF and map servers are redundant in each data center site, creating quad-redundancy.

Because the nature of a secure ESInet is that it is private, the network must control the edges with the BCF; this requires DNS and other functionality often provided by Internet service providers that the NG9-1-1 system must support standalone. Moreover, the ESInet is hungry for bandwidth, which is best served with fiber optics in a ring as noted throughout this book. At the onset, a lot of areas did not and still do not have adequate bandwidth and or adequate public or private fiber. To boost infrastructure in the United States, the federal government offered grants called Broadband Technology Opportunity Programs (BTOPs), and those who applied needed to match funds requested by 20%. Some regulatory bodies did not allow for the use of private fiber, or at least not for some 9-1-1 SSPs.

Interfacing ancillary systems like CADs can be expensive because of the proprietary nature of CAD systems. The vendors' goal is to "lock in" customers with property databases that make moving the data prohibitively expensive. Interfacing with non-IP-based CADs can be accomplished but has limited benefit. All of this adds significant cost and or time for the project.

Since NG9-1-1 requires contracts and legal expertise, legal fees add to the cost.

"Time is money," Benjamin Franklin told a young apprentice[1]. To build a secure ESInet with POIs for direct OSP connection takes time. Hofstadter's Law[2] states: "*It always takes longer than you expect, even when you take into account Hofstadter's Law.*" During the implementation of the NG9-1-1 project, the existing E9-1-1 network must continue to be paid for as well as likely upfront payments to the NG9-1-1 RFP developer, the selected 9-1-1 SSP, the SI, and other vendors. OSPs react in different time intervals. Some cooperate, some procrastinate, and some OSPs resist, especially if they are being displaced as the 9-1-1 SSP. It always takes longer than the original plan, and the authors' experience is to calculate the required time and multiply by two. Early RFP bidders made commitments for deployment based on technical time to deploy as if they were replacing E9-1-1 only

Some deployments claiming to be NG9-1-1, in fact, turn out to be hybrids, since the 9-1-1 calls still come through the SPOF legacy SR. Those SR charges were supposed to be replaced by the NG9-1-1 core data center costs. Agencies are paying for the old 9-1-1 SSP and a new "behind the scenes" 9-1-1 SSP. This approach can take less time and perhaps appear to entail a smaller price tag, but it is a myth that this design is i3-compliant. It lets the OSPs maintain a lack of commitment to connecting directly with the NG9-1-1 network, but in no way is this more reliable or efficient or less costly to the public safety agencies. In the PBX world this was

1. "Advice to a Young Tradesman," by Benjamin Franklin, an essay that appeared in George Fisher's 1748 book
2. Gödel, E., *Bach: An Eternal Golden Braid (20th anniversary ed.)*, 1999, p. 152.

called "IP behind the switch," and that phrase applies to these configurations, as they violate the redundancy and geographic diversity design fundamentals.

The last myth is that there is lots of federal money to support the transition to NG9-1-1. Many myths have an element of truth, and this is one such example. In the United States there is federal grant money available, but it requires matching funds. There is also a limit as to how much money is available and what it can be spent on. In CSI, the solution around the 20%[3] matching funds requirement was alleviated when a group of consultants and one university laboratory volunteered labor hours in the application equal to the value of the 20% match. To be a part of history and help part of the United States where some counties had no 9-1-1 service, it was worth the commitment.

Funding comes down to how much of the budget and reserve funds the local public safety agency can allocate to the project. The answer depends on scale. Larger agencies may be able to self-fund the project, but this is still expensive, and that is why state and regional projects are so attractive: By combining resources or agencies, such partnerships can create economies of scale and cost savings, since they spread the fixed and variable costs over several agencies and process the same number of calls with consolidated PSAPs.

And that leads to a myth within a myth: consolidation. There are physical and logical consolidations. CSI, for example, never consolidated a single physical PSAP for nearly five years after the conversion to NG9-1-1. The counties formed a legal entity called CSI and allowed the call takers in any PSAP to answer the calls for any other PSAP if the primary and alternate PSAPs were not available or had insufficient personnel. They were fortunate to have common radio systems so they could dispatch for each other. As mentioned in Chapter 10, common radio is a feature of FirstNet, and it can eventually help all agencies along this journey. CSI did not consolidate its staff to a single payroll. If that had been done and a single CSI PSAP team began serving much of the southern third of Illinois, then some centers could have shut down in low hours or on low days. Staffing mostly three- to five-seat PSAPs could have reduced overall manpower or at least reduced holiday and overtime costs. NG9-1-1 standards do not offer guidance on staffing, consolidation, or manpower. Standards allow what an area wants to financially support. On the other hand, the persons answering all the NG9-1-1 calls are and were seasoned professionals who had worked on common processes and who were managing the business as a team. This is not always the case with public safety where elected officials feel they need control their "turf."

12.2 Sources of Funds

12.2.1 Surcharge

Surcharge is a euphuism for a 9-1-1 tax on a phone bill. Surcharge exists for wireline, wireless, and VoIP service although the rate, remittance level, and receiving entities vary by state and locality. Wireline surcharge—typically determined by a local ordinance, city, or county—ranges from $.25 to $5.00 a month, and, in some cases,

3. The matching fund rate can vary from grant to grant. In the CSI example, the rate at that time was 20%.

it is a percentage of service charges from .0075 to 12%. Some cities or states require voter approval to raise the 9-1-1 surcharge. These initiatives are seldom approved by voters. Wireline surcharge is a rapidly declining revenue stream, although it was formerly the primary funding mechanism of E91-1. Wireline surcharge was typically remitted directly to the public safety agency but over the past five years more states are taking control of the wireline surcharge and distributing it to the local agencies—after, of course, the state takes a fee for handling the money. This action is sometimes used as "a money grab" or redirection of funds for other purposes.

The wireless surcharge rate is typically done on a statewide basis and ranges from $.25 to around $1.65 a month; the City of Chicago's wireless surcharge is a higher $5.00. Surcharge remittance for wireless service is typically to the state as described above for wireline service. While wireline surcharge is remitted for the connections at the service address, the wireless surcharge is submitted to the billing address. The problem this injects into the funding models for public safety is that large corporate accounts bill wireless service to their headquarters' locations, and thus wireless service that is used in one state may have the surcharge for that device remitted to another state. When that occurs, it deprives the agency that answers the emergency calls but receives no compensation for the device. Of course, other agencies receive a surcharge windfall from merely having wireless bills sent to their jurisdiction.

VoIP surcharge often matches either wireless or wireline if a VoIP surcharge exists. The remittance is similar to wireless in that it applies to the billing address, not the servicing public safety agency location.

Surcharge struggles to provide adequate funding for public safety. Declining wireline revenue has not been covered 100% by increases in wireless or VoIP surcharge. All of this creates funding instability and an increasingly longer time before the public agencies see the money. That means that the vendors they have contracts with are seeing late payments; this snowballs to disputes, late fees, and a list of grievances. Some cities and states, Nevada specifically, have shifted from a surcharge to a sales tax or property tax to fund 9-1-1.

The FirstNet management and funding model discussed in Chapter 10 and supported by congressional funding and mandates—paired with a firm budget process and accountability—resolves much of the piecemeal and frustrating approaches to funding NG9-1-1.

In NG9-1-1 applications to states that require regulatory approval, a five-year funding plan is often a mandatory part of each public safety agency's application. Sources of funds, funds on hand, and how funding is managed are essential to NG9-1-1 implementation approval.

12.2.2 Grants

Grants represent money for public safety agencies that do not have to be repaid. Grants for NG9-1-1 are federal tax dollars made available for agencies that qualify and can conform to the requirements. Agencies seeking grants should visit: https://www.911.gov/project_911grantprogram.html

From the 911.gov website:

> The 911 Grant Program provides grant funding to support state and local efforts to deliver optimal 911 services. The funding may be used for the implementation and operation of 911 services, E911 services, migration to an IP-enabled emergency network, and [/or] adoption and operation of Next Generation 911 services and applications0

The website goes on to say,

> The 911 Grant Program requires cost sharing, also known as a matching funds requirement. The federal share of the total cost of any activity carried out under the Grant Program may not exceed 60% of the eligible cost of carrying out grant activities.

12.2.3 OSP Funds

The love of money is the root all evil.
—1 Timothy 6:10 *King James Bible*

Lack of Money is the root of all evil.
—George Bernard Shaw

The OSPs are a component of the funding formula and can add to the cost of migrating to NG9-1-1. In the telco world as OSPs exchange traffic, there are charges to originate and terminate traffic. This is a complex process known in the industry as carrier access. This remnant of the early days of interconnection appears in the 9-1-1 world from some OSPs. 911 SSPs often charge OSPs to terminate 9-1-1 traffic into the E9-1-1 SRs. CLECs are accustomed to this arrangement. Some OSPs expect to be paid to deliver 9-1-1 calls and provide location information.

ILECs tend to be revenue-centric organizations, and independent telcos have been subsidized, so they usually expect to get compensated for any modification. 9-1-1 SSPs that are large telcos factor compensation based on the other relationships they have with the OSPs in terms of interconnection. 9-1-1 SSPs that are not large telcos can face significant costs and potentially moderate revenue when carrier access type charges are applied. The goal remains better service for customers end to end, so regulatory accommodations need to be made to facilitate service and cost sharing. Contracts with firm SLAs for nonperformance are always required—generally ahead of project approvals. Without a deep dive into OSP interconnection, independent 9-1-1 SSPs should probably take the position of: "I won't charge you to terminate 9-1-1 calls and you won't charge me for the trunks." The position the authors took in these negotiations was that the person calling 9-1-1 was a customer of the OSP. The OSP is required to provide customers access to 9-1-1. The 9-1-1 SSP is the agent of public safety agency that in most states designates the location of the circuit termination. The bottom line: We take your 9-1-1 calls for free, and you provide the 9-1-1 calls for free. Of course, there are costs associated with this activity, but each party accepts the costs to provide and receive 9-1-1 calls.

12.3 Successful Funding

Successful approaches to funding include CSI, which received BTOPs and other community-oriented policing services (COPS) federal grants that were available during the Obama administration. The CSI project was facilitated by Clearwave Communications, a CLEC receiving over $45 million to build a fiber-optic network across southern Illinois counties. The Clearwave grant was rejected until public safety was included in the grant application. While the individual agencies did have to match on certain grants, Pat Lustig and Ken Smith finessed the grant process to get probably the "deal of the millennium" to build their NG9-1-1 system. Finesse included volunteer labor to reach what was then a 20% match requirement. A similar story unfolded in northern Illinois in the Northern Illinois Next Generation Alliance (NINGA) process, and significant funds were granted to a consortium commonly known as Northern Illinois University (NIU) and iFiber. This effort was a success for NIU and iFiber, but a public safety NG9-1-1 failure due to the reluctance of the Illinois Commerce Commission (ICC) and Illinois State Police Regulators, the objection of the incumbent 9-1-1 SSP, and leadership within NIU and iFiber. The issue was control. The NIU and iFiber consortium did not want to become a CLEC, regulated by the ICC. The ESInet needed the fiber network for the most cost-effective connectivity for public safety. In addition, data center owners and operators wanted a commitment that they could indeed "kick out" public safety if they decided to sell their data centers or went out-of-business. NG9-1-1 needed a default network. Only telcos and the state itself owned enough alternate fiber or similar connectivity. The state fiber network, Illinois Century Network (ICN), did not want to become a regulated CLEC. The telco bandwidth was high-cost and not always diverse or available to each PSAP. The liability could be mitigated with tariffs, and the ESInet provider did not want to file tariffs. Contracted SLAs with the 9-1-1 SSP were tightly controlling the service quality, but the state listened to the incumbent 9-1-1 SSPs and convinced the administrative law judge (ALJ) that there could be risks to public safety. Fast-forward to three more years of waiting, a new 9-1-1 SSP, the same network and plans, and the approvals were given. What really happened? Nothing except legal maneuvering and delay, but that cost time and, of course, money. In the meantime, some small businesses went out of business, and more lives were impacted due to delays and failing infrastructure. Nevertheless, an abbreviated version of NG9-1-1 eventually is moving forward with new faces and fresh challenges.

12.4 Summary

The best way to ease and contain costs is to achieve economies of scale by having a large number of agencies/PSAPs or a large number of workstations using the same NG9-1-1 core. In other words, increase the number of 9-1-1 calls processed by the expensive part of NG9-1-1, the core, and ESInet. States that are moving to provide statewide systems should achieve economies of scale if the resources and design are configured and concentrated efficiently. The public safety agencies are often part of a fiefdom in local government. The sheriff and /or police chiefs are powerful agents who want to—and in many areas should—control almost all most areas of public safety. There is nothing nefarious in the desire for control, but for midsize to smaller public safety agencies, the concept of "we need to own core and control it 100%" is a significant factor that is holding back the wide deployment of the NG9-1-1 network. The cost of NG9-1-1's core alone excludes many public safety agencies from moving to NG9-1-1 if they remain independent. These agencies are determined to hold onto control, and there is much resistance to consolidation in addition to the general resistance and fear of change. Consolidation would create economies of scale to reduce the total cost, but cooperation is necessary for NG9-1-1 success. As we close this chapter, we are reminded that there were three factors and, in fact, requirements in gaining approval for migration to NG9-1-1 in the eyes of the regulators we dealt with—and in any business case model. They are technical, managerial, and financial. First, do you have the technical skills proven to make the transition and implement the work successfully. Second, do you possess the managerial expertise and requisite skill sets to perform the duties across the board. And finally, can you prove you can fund the project, with financial depth and backing and a proven track record for managing cash flow and reaching the end of the project. With the inherent delays in the move to NG9-1-1, vendors, systems integrators, 9-1-1 SSPs, and public safety agencies struggle more with the last requirement.

CHAPTER 13

International Perspective on Emergency Services

13.1 International Demand for Emergency Services

Sister Theresia of the White Sisters Order of Kenya, Africa, shared her story, which is at the heart of the mission for basic emergency services in some parts of the world today. Residents in the jungle village 50 kilometers from her clinic/field hospital were carrying the sick and injured on litters[1], with up to four human carriers per patient to be triaged as recently as two years ago. She was soliciting funds for an ambulance to transport patients faster and to provide emergency care along the way. The villagers were not able to call for help, and many were dying because of delayed care.

When the author was teaching and mentoring student projects for IIT, the students were from all over the world including the United States. One young man from the Congo was returning to his country to set up communications systems, and he wanted to learn about network management, emergency services, and disaster recovery. His home community had hope of a better outcome in their daily lives upon his return. We discussed 9-1-1 services and the types of issues that the Congo faced. Student projects were utilized to study disaster recovery principles and to practice with IIT's VoIP laboratory to determine how they could improve the whole infrastructure to make it more resilient, reliable, and secure. Students from around the world attend IIT, get their master's degrees, and are at the pinnacle of knowledge in new IP-based technology. However, these students have no practical experience with the oldest of communications technologies that needed to be transitioned to the future. The whole concept of being able to call an operator or a call taker with a dedicated number for emergency access was simply not in the vernacular of students from many countries. Other students from Europe and

1. The litter is a class of wheelless vehicles, a type of human-powered transport, for the transport of persons. Smaller litters may take the form of open chairs or beds carried by two or more carriers, some being enclosed for protection from the elements. Larger litters, for example, those of the Chinese emperors, may resemble small rooms upon a platform borne upon the shoulders of a dozen or more people. To most efficiently carry a litter, porters either place the carrying poles directly upon their shoulders or use a yoke to transfer the load from the carrying poles to the shoulders. (Source: Wikipedia.)

technology-advanced countries had an advanced baseline of technology and access. Europeans had the genesis of their 1-1-2 systems and were using advanced wireless and PCS technology. Most countries' communications were driven top-down by the postal, telegraph, and telephone service government organizations that owned and operated the wireline and wireless networks, although competition had begun to enter the picture.

Every year more and more students started emergency services projects in the IIT lab because of vendor equipment donations. Student classes included projects to learn about the technology hands-on and build upon the knowledge through semester projects. Some vendors brought new projects and products into the lab for further development and testing such as Radio over IP and BCFs. Students reviewed standards and learned about wireless, VoIP, and PBXs, finding freeware and creating solutions each semester. Annual conferences at IIT included a track on 9-1-1 and eventually morphed to NG9-1-1 as the interests merged with the evolution of the standards. This chapter explores a sample of international communications standards and technology for emerging and advanced emergency services.

13.2 International Emergency Calling Patterns

The need for emergency services is universal. From their earliest years, we drill into our children the importance of calling for emergency assistance. NENA has expanded its mission and outreach to include like-minded countries and organizations around the world, partnering and developing standards that are not simply North American–centric concepts. The key is developing and contributing to standards for advanced and—as denoted by this book—next-generation emergency services. Others around the world are aware of NENA's and APCO's plans and are finding their own paths forward to improve emergency service goals using advances in underlying technologies to achieve national objectives.

A simple numbering scheme to reach emergency services is a common approach, yet an icon or red button on a device may facilitate the equivalents of an SOS[2] or Morse Code signal for emergency assistance in others' visions. As world travelers, workers, and consultants have learned, there is no simple numbering scheme, and travelers need to dial the full local numbers for emergency assistance. The ubiquitous "0" operator may still exist. An Internet search on a smart phone, if a connection is available and no language barrier exists, may be a means of getting assistance. From a practical standpoint, however, it is important to research and have emergency assistance information at hand before traveling around the world.

Before you need help, verify how to get it. Do not assume that what works from your landline—if you still have one—or your personal wireless device works

2. SOS is a Morse code distress signal, used internationally, that was originally established for maritime use. In formal notation SOS is written with an overscore line to indicate that the Morse code equivalents for the individual letters of "SOS" are transmitted as an unbroken sequence of three dots, with no spaces between the letters. In International Morse Code three dots form the letter "S" and three dashes make the letter "O," so "S O S" became a common way to remember the order of the dots and dashes. (*Source:* Wikipedia.)

Table 13.1 Sample Emergency and Short Codes Worldwide

Country	*Short Code*
Australia	0-0-0
India	1-0-8
Iran (Police)	1-1-0
New Zealand	1-1-1
Pakistan	1-1-2-2
European Union and Others	1-1-2
Parts of Asia and Jamaica	1-1-9
China	1-2-0
North America and Philippines	9-1-1
United Kingdom and others	9-9-9
Worldwide: Check before you develop or travel.	Local numbering schemes with full telephone numbers for access

the same when you are on the road. If it is important, test for access, especially for older, more infirm people living alone; for younger children and childcare givers who may be responsible for children on field trips; and in rural remote areas such as camps. If a college student is abroad this becomes "need-to-know." The various participants in moving NG 9-1-1 forward need to be mindful that the market is worldwide and, accordingly, plan to design and test the systems with a worldwide numbering scheme and user community in mind.

13.3 Providing Emergency Services

It takes time to address the worldwide requirements, but rest assured that many countries and international universities attend the same conferences and read the same standards, and this is not a case where one company is dictating standards and the rest of the world is expected to follow. Participation is open to everyone now, and leadership in emerging standards and access ideas comes from around the world. The first edition of this book includes a European sample.

13.3.1 European Perspective

The heart and mission of emergency services is aiding the public safety personnel who save lives everyday around the world. With that in mind, we interviewed one of the leading ICE co-chairmen and sponsors of advanced capabilities across Europe, Wolfgang Kampichler, who, at the time, was the principal scientist at Frequentis AG. We met while planning for ICE 5, the first NENA-sponsored event hosted at the IIT RTCL. ICE is not simply a United States–supported event; it is often led by experts in next-generation emergency services from around the world. Kampichler shared the status in Europe and educated us particularly with respect to the European market. In the past there was a plethora of access numbers for emergency services in Europe (e.g., 15, 17, 18, 110, 113, 115, 122, 133, and 144). Through standards, the numbering plan moving forward is 1-1-2 as the single access number.

Where the calls first route varies by country, with some to police, others to fire, and then others to a combination of police and fire answering locations. The European Electronics Communication Code (EECC) was under revision to include requirements for reverse 1-1-2,[3] privacy, accessibility, and location codes. There are or were a lot of PSAPs in each country when Kampichler provided his statistics in 2018 (i.e., 158 PSAPs in Austria alone). There were three basic pieces to the equation: the calling originating network, which we call access carriers or OSPs in the United States; the emergency call service providers (ECSPs), which we tend to call the 9-1-1 SSPs that provide the mapping and routing function to reach emergency services; and the PSAP service providers (PSPs) or IP network providers connecting the PSAPs. In the European vision the mapping systems of the future migrate to solutions based on caller location, which is the civic address, or the geolocation augmented by media and data. That future vision aligns with the vision of NG9-1-1 in the United States.

The PSAPs will be connected by an ESInet or series of interconnected ESInets where calls can be directed or redirected to different agencies as needed. EENA addresses the process in stages. First, PSAPs are upgraded and given full access to each other and a means to send calls freely as needed to the right first responders. The parties in charge can understand these improvements and plan for and fund these changes. The concept of a multiagency or "statewide" team of PSAPs has emerged.

13.3.2 International Location Services

A European Country (EC) mandate Kampichler refers to as M/493 set the path toward improved location services. ESTI ES 203.1.78 for Access Architecture supports the location process. The reference standard includes those for protocols and signaling: the European Telecommunications Standards Institute (ETSI),[4] ES 203 183, and Signaling SIP, and location query, retrieval updates using HELD or MLP. The specifications allowed for national customization but clearly included many of the standards and FEs mentioned in Chapter 3, a service URN, PIDF-LO, and a nationally provided PoP,[5] which is their reference to a firewall for secure access. As Kampichler mentioned, the management in charge could more easily recognize and fund solutions for access and location capabilities. Access is easier to comprehend (i.e., fund) than the "architecture/systems" in the middle that we tend to think of as the core services of NG9-1-1, the NGCS.

3. Reverse 1-1-2 is similar to the United States' reverse 9-1-1. Parts of the United States have reverse 9-1-1 where the police typically inform the residents who subscribe of ongoing events in a community such as burglaries in progress, road closings, or major threats. This may take the form of an email, phone call, or text alert.
4. ETSI is a leading standardization organization for ICT standards fulfilling European and global markets.
5. Exchange Best Practices: Secure POP and IMAP Access, May 24, 2016 by Paul Cunningham 6 Comments, "A newly installed Exchange 2010 or later server has the POP and IMAP services disabled. The POP/IMAP settings for the server are also configured with secure default settings, so that if you were to start the services they would operate in a secure fashion by default. The main concern with POP/IMAP security is the login process. POP/IMAP protocols allow login over unencrypted connections, transmitting login credentials across the network in clear text. By requiring secure logins on the Exchange server, the credentials are passed over an encrypted connection, protecting them in transit."

As to core services, while they are essential for end-to-end- success—as we note repeatedly in this book—it is the last area to be required and/or funded across Europe. ETSI SC EMTEL work items included TS 103, 479, and TR 103 and 480 to achieve what we tend to call NG9-1-1 core services. The terms for the FEs are the same as defined in Chapter 3 for infrastructure, the BCF, LIS, ESRP, ECRF, and i3PSAPs with their interfaces. EENA NG112 LTD was fast becoming an ETSI standard.

13.3.3 International Advanced Mobile Location

In the United Kingdom, British Telecom (BT) began advanced mobile location (AML) research in connection with emergency services. This initiative was supported by EENA, was pushed by Google Android products, and then supported by iOS products. Standards include ETSI TR 103 393, plus new EMTEL WI. AMS is an, "automated location conveyance when an emergency call was initiated." ETSI RFC 7378, "Silent activation of different location methods were used with sensor fusion and the messages sent via SMS and/or HTTP PUSH." Basically the PSAP receives the location information sent from the aggregating services provider (ASP) to the PSAP service provider (PSP) to the correct PSAP. Using short message service (SMS) and a data exchange format developed for this purpose, the location arrives with the call. The AML and the Pan-European Emergency Mobile Emergency App (PEMEA)[6] support advanced mobile location services in Europe and have near-term international implications and advantages. PEMEA technology is used for moving entities and location to PSAPs; these technologies are not apps in the common use of the term, but requirements for the device manufacturers to embed in their technologies.

13.3.3.1 Telematics

Telematics in Europe generally operate differently than in most of the United States. Cars are required to have emergency services capabilities. The call does not go to a private agency such as the car manufacturers' call centers as in the United States, but instead it is sent via wireless GSM phone network to the appropriate PSAP. That PSAP may be a single-purpose PSAP such as a police station or fire brigade. From that point, as needed, the calls are transferred to the correct agency. That could be a combination of ambulance and or fire brigade in the case of a car or train accident, for example. The European model seems to be better at requiring the car to send an "SOS" or request for assistance right away, automatically getting help without many transfers.

6. The PEMEA architecture allows emergency apps to interconnect so that a citizen in distress can use any emergency app anywhere across Europe. The PEMEA architecture itself is not new; it has already progressed through ETSI as technical specification TS 103 478, making it a European standard. However, now the focus is on actual real-world deployments across a range of regions and countries throughout the EU." (*Source:* Beta 80 International.)

13.3.4 International Text to 1-1-2

Texting to emergency services is in progress in Europe, though there are different types of text solutions in place, none of which are truly standards-based: (1) SMS text, (2) the number being texted from is not a real number and thus calling back the calling party is not an option, and (3) no solution. Texting today varies but most of the text solutions are one-way texting with the PSAP unable to acknowledge or ask questions. The texter does not know if the message is received or if help will arrive. The calls arrive at a single agency, such as police default in Austria and fire brigade in Germany, and the calls are transferred to the correct ambulance, fire, or police depending on the message and whether sufficient information arrives to identify the problem and the location of the incident. Texting is an important topic internationally, especially for persons who are hearing-impaired. Next-generation emergency services solutions offer hope for improvements. The first step would be the capability of two-way chat and efficient routing. Regulations for this area would be helpful to Europe and internationally.

13.3.4.1 Commentary on Interim and Transition Issues

As people travel throughout Europe, the numbers they are accustomed to dialing for emergency services vary. In many countries a call for "9-1-1" is accepted and routed to the nearest default PSAP even if the dialing code is 1-1-2 in that country. Not all countries have the same universal access code. These are recognized issues, and regulations need to be promulgated to adhere to standards as they evolve. Ultimately, and this is true in North America and elsewhere, true next-generation solutions include icons on the phone for police, fire, and ambulance, and the translations to the proper underlying routing based on the location of the caller takes care of the unique differences across the networks.

13.3.4.2 Speed of Migration, Transition Issues, Barriers to Success

Items of importance to the dialog in Europe are language identification of caller, location of caller, database, country-to-country ESInet to ESInet, and legacy-to-next generation issues. The same set of overall goals address future requirements and next steps. These are described as follows:

1. Barriers appear to be a lack of mandates, especially as related to the core network's migration to a next-generation standard.
2. Access carriers and access in general has not been addressed for the migration to a NG core diverse architecture since no one has that in place.
3. ESInets are not being addressed today in Europe.
4. In terms of the language of callers, in Europe, when an emergency call is answered, the agency tends to get a person online who can communicate or the call is transferred as needed to a language specialist, which loses time and response. Adding the native language of the caller into the handset would improve routing to the correct PSAP.
5. It appears that Europe is working on PSAP replacements but not necessarily next-generation technology end-to-end.

13.3.4.3 Operations Post-Implementation, Monitoring, Problem Resolution, Reporting Requirements

Operations is another topic that applies universally but includes enterprises with direct connectivity, vertical location, and a myriad of issues of common interest.

13.3.5 Legal Regulatory Issues Across Europe

13.3.5.1 Persons with Disabilities

Reference the EECC. It is mandatory that the public and providers look at regulations regarding persons with disabilities and their access to emergency services. Offerings must include handset-derived location (i.e., the carrier handset must provide location as part of the code).

13.3.5.2 Social Media

When asked about the use of social media in emergency services call centers, Kampichler said that public safety greatly relies on the use of social media to identify big issues and to alert the public to major events in progress. This is a general statement, not one about next-generation or interim emergency services as provided today or tomorrow. Integration of social media to PSAPs is not recommended as a direct means to report or get notified about a major event affecting a large area. The idea of public safety apps does not seem to be in use in European PSAPs as compared to the United States.

Examples demonstrating where the public benefits from social media include earthquakes, tornadoes, road blocking, mass shooting events, or bombings; in such situations, the public can see something and say something that reaches friends, families, and many general contacts. At the actual site of a major event, the callers need full access to emergency services with the specific purpose of reporting the event and getting immediate help. In that case, the notification should be made by calling the emergency services agency specifically equipped to dispatch the right first responders to the scene. Videos and other photos may be helpful; in next-generation emergency services, the standards include the transport of video, photos, and data to the PSAP.

13.3.5.3 General or Focused Overload

One of the risks involved in today's older embedded emergency services networks is the possibility that the networks can overload. A general or focused overload such as a DDoS attack can compromise first responders' ability to receive calls in a massive event. Modern next-generation services networks are the very best means of protecting the PSAPs and first responders from overload. The use of the BCF, for example, if properly installed and managed—and the importance of these items cannot be stressed enough—provides rapid alerts if an overload is beginning and allows the PSAP agencies to open up emergency management centers and back up locations for answering calls. These networks are designed to withstand the attacks and overloads whether they are purposeful or the result of natural events like weather or football (or soccer for American readers); such events can cause emergency services to get overloaded. Core next-generation networks are not being funded in Europe. This is a gaping hole in public policy, and education is necessary to overcome this inertia in regulation and implementation.

13.3.5.4 Legal and Regulatory

Legal and regulatory issues vary across countries and regions as do the bodies that promulgate the requirements. European regulators have standards-setting bodies that promulgate many of the emergency services–related rules. Implementation and time frames for conformance vary significantly. Until and unless there is public demand for the implementations coupled with a willingness to pay, the situation may only improve incrementally. The PSAP upgrades to next-generation technology and the location requirements are being addressed, because the regulators and the public can understand location services, and the regulators and PSAPs understand when their equipment is not functioning and needs to be replaced. The core of the next-generation emergency services network requirements—which includes the design of a diverse ESInet, diverse locations for housing fully duplicated next-generation equipment for routing and data management, centralized logging and recording, failover and accurate PSAP routing where all the relevant information stays with the call—is the province of either the state/country (approximately 20%) or private companies (approximately 80%); telcos often provide that service, and they are not under any regulation requiring them to migrate from the approximately 50-year-old architecture and technology of emergency services to the next-generation capabilities. The development of ESInets is not taking place in Europe today. Citizens and visitors to each country are at risk of not getting rapid help. Education is needed, and the various entities participating in standards development, including systems integrators and systems developers, need to educate the public and legislators. These are comments derived from conversations with Wolfgang Kampichler in August 2018.

13.3.5.5 Privacy

Embedded in the regulations related to emergency services in Europe is a statement releasing the PSAP agencies from protecting the privacy of the person requesting emergency services. The name, location, and all identification needed to find the person and/or place where help is needed supersede all expectations of privacy. On the other hand, more personal information comes from the newer networks and devices, including voice, data, video, photos, and text. This information's mandatory location in the handset means that the caller does not control location information going forward. If a car is in an accident, and telematics sends the call to the PSAP, the information may include the number of persons in the car, how fast the person was driving, and any health conditions that can be derived from the vehicle. The PSAP agency uses that information to get first responders to the scene and have adequate information to assist in saving lives. The information is to be stored according to prescribed practices and protected from outside intrusion and access. The information is to be destroyed when appropriate as well.

13.3.5.6 Security and Cybersecurity

Many issues related to security and cyber security seem to be the same in Europe as the United States. The introduction of IP technology coupled with the addition of the BCF in the next-generation architecture is key to preventing cybersecurity breaches inside the emergency services networks.

There are best practices for access security in the PSAPs. This includes physical and logical security. Monitoring at NOCs is crucial once a network is built using standards. Login management and password management once someone is in the system is critical. Physical access and door codes and access to rooms where equipment resides is crucial. The entrance of the public and/or vendors and media into important areas needs to be supervised. The network includes remote transport sites, public and private power stations, and many other locations that are known today and that may change with recent technology.

Review Sections 3.4, 5.7, 6.8, 6.11, and 7.3 that include security, and pay special attention to the requirements for ongoing active operations, including the monitoring solutions and tools needed to protect the emerging next-generation emergency networks.

13.3.5.7 Differences from the North American Model

There are common issues international public safety agencies share. In the United States, as noted in Chapters 11 and 12, there are multiple and variable regulatory agencies that have oversight at the federal level, where persons with disabilities are represented by legislation from Congress; the FCC regulates wireless QoS and requires reports from OSPs or access carriers where emergency services calls originate and terminate to PSAPs. The PSAPs are regulated by the states, usually through state commissions. In some cases, strict regulations demand that a 9-1-1 SSP be certified and approved to provide emergency services within an area of jurisdiction. In other cases, only a state-certified OSP or access carrier may provide underlying network connectivity (i.e., switching, transport, and routing of the emergency services calls to the PSAPs). No two countries or even regions of countries internationally have a common form of regulation, although, as in the United States, some regulation appears to be present where emergency services are provided to the public. Research must be done for each next-generation project. Funding to match those regulations varies as well. Kampichler shared his experience operating in Europe and participating in standards bodies worldwide: "It is clear that around the world, not everyone has access to emergency services, and there is no single common denominator that makes it easy or mandatory to move forward expeditiously, and in a manner that meets next generation standards crossing country, state, regional and local boundaries."

13.4 International Forums and Standards Bodies

ETSI is an important standards body across the European Union (EU). "ETSI is an independent, not-for-profit, standards organization in the information and communications technology industry fulfilling European and global market needs." It is based in Sophia-Antipolis, France. ETSI supports the development and testing of globally applicable technical standards for ICT-enabled systems, applications and services that are widely deployed across all sectors of industry and society. ETSI authorizes and supports interoperability plug tests similar to the ICE events supported in the United States by NENA. By extension, it supports conformance testing, which is not yet in place in the United States. An ETSI Specialized Task Force (STF) develops an abstract testing and control notation (TTCN) test suite to cover 1-1-2 core services interfaces. It uses open-source TTCN tools to accomplish its mission.

NENA and APCO are international standards organizations. In Europe NENA is known as EENA. As varied as numbering plans are around the world, there are often posts and telecommunications governmental organizations that drive or adapt standards for their own countries. The IETF, IEEE, and many other organizations promulgate standards that include a worldwide audience. There is no longer a strong desire to promulgate different basic standards for emergency communications when it comes to the concrete reasons emergency services standards were developed. IP technology is not the domain of a single country or manufacturer.

13.5 International Summary

Around the world, calls originate from varying networks and services providers. How the calls are routed today depends greatly on the PSAP or call-answering model in place from a national, regional, or county level. In some cases, the implementer of next-generation emergency services is starting from a "green field of operations" where nothing concrete exists. Most of the time, in developed countries, there is a lot of sophisticated infrastructure in place, but it is often based on older technology. Standards are geared at providing new approaches to an age-old challenge. Common methods can lead to more rapid progress. As you approach the future, consider different types of media, location-based policies and call routing, and security and privacy constraints and align your goals with standards bodies and forums that are proven and consistent and that operate from an international perspective around the world.

Selected Bibliography

Conversation in 2018 with Wolfgang Kampichler of Frequentis; he is a member of most European standards bodies and a key coleader of NENA ICE testing events.

Appendix A
Method of Procedure Template for 9-1-1 Cutover, Upgrade, and Changes

Date prepared: __________
MOP number: __________
Planning or engineering order number: __________

Note: In the event the planned activity causes a service interruption, the MOP leader shall do the following:

1. Cease all work immediately.
2. Notify all pre-identified personnel responsible for the area(s) impacted.
3. Determine next steps and proceed with approval.
4. Enter the name of centers, organizations, and personnel in priority order to be notified if service is at risk or degraded.

	Name/Center	*Phone*	*Text*	*Conference Bridge*
1				
2				
3				

Add as many lines as needed.

Provide a list for each vendor or supplier company involved in the MOP.

Installation Supplier Personnel Working This MOP					
Name/ Company	*Skill Level*	*Emergency Number*	*Alternate Contact*	*Skill Level*	*Emergency Number*

Add as many lines as needed.

The supplier must not deviate from the signed approved MOP, unless authorized by the 9-1-1 SSP.

MOP Approval

Title	*Print Name*	*Mobile Phone*	*Signature*	*Date*
MOP leader				
9-1-1 SSP CTO				
Installer Supplier CTO				
9-1-1 SSP Equipment Engineer				
Power Representative				
9-1-1 SSP NOC Director				
Others as Appropriate				

Add as many lines as needed.

MOP name and number: ___________________________

Note: Add as many pages as required.

Advanced testing required or dry run to determine time and tools

Yes____ No____

Add names of personnel for dry run and date and time. Include steps to verify success.

Changes from dry run implemented

Yes____ No____

A checklist for preparedness is to be answered affirmatively by each center and participant before the MOP proceeds. A conference call may be the quickest way to ask these critical questions and document affirmative responses.

1. Does the team have approved secure building access as required?
2. Is a company ID required?
3. Does each team member understand the goal?
4. Does each team member understand reliability requirements?
5. Is the sequence of events clear?
6. Has everyone been notified: customers, commission staff, NOC personnel for each company performing the duties, tiers I through III?
7. Is monitoring in place across the area being impacted by the change?
8. Is the back-out and recovery plan documented and understood?
9. Does each team member know the maintenance window for the work?
10. Is each team member trained, and do members have the proper tools to perform the work safely?
11. Is the work plan clear and comprehensive?
12. Do team members know their primary escalation contact 24 by 7 by 365?
13. Is there a risk that duplex operations could be impacted?
14. Has each party signed off as required well before work begins?

15. Is the person required to make reports to commissions and customers aware of how to reach the correct parties immediately and provide notification and ongoing updates as required?
16. Should commission staff be notified in advance of the event?

Signatures	*Names*	*Dates*
9-1-1 SSP signature		
Installer vendor signature		
NOC manager signature		
Tier II technical support signature		

Add as many lines as needed.

Describe the plan including the maintenance window for safe stop
The detailed steps of work shall be listed and completed sequentially.

Responsibility				*Describe Steps*	*Completed*			
Step No.	*Supplier(s)*	*OSP(s)*	*9-1-1 SSP(s)*	*Description of Work and Critical Activities*	*Date*	*Time*	*Supplier(s)*	*9-1-1 SSP(s)*
1.								
2.								
3.								
4.								

Add as many lines as needed.

Maintenance Window Agreed Upon: ____________________ include time zones.

Attach any pertinent drawings, equipment details and processes. Indicate the total number of pages.

About the Authors

Barbara Kemp is a partner in Assure911, LLC. Her mission is to assure that clients have the ability to build, monitor, and manage their emerging 9-1-1 networks end to end. She is a 1983 graduate of Indiana University and a 1992 graduate of the Wabash College Executive Management Program. She taught network management at the Illinois Institute of Technology (IIT) as an adjunct professor and mentored graduate students in IIT Real-Time Communications Laboratory projects.

As project manager for Cable and Wireless Caribbean operations, Barbara consolidated NOCs and built a new NOC for 13 Caribbean nations. Barbara managed a wide range of IP- and circuit-switched consulting projects for telcos and cable TV companies nationally and internationally. Barbara designed, built, and managed NOCs, including AT&T's Midwest Reliability Center and ICG Telecommunication's Service Reliability Center as senior vice-president for Service Reliability, Quality Assurance, and Security. The NOCs included emergency services problem resolution, business processes, systems integration, and security. She managed an emergency services technical resolution task force after the City of Chicago embarked on its city-wide Operations Emergency Management Center (OEMC).

In addition, Barbara was the professional services engineering and design consultant for a next-generation 9-1-1 project for the CSI from planning through implementation. She also developed a NG9-1-1 plan, supporting the Northern Illinois Next Generation Alliance (NINGA).

Bart Lovett, a partner with OnPoint, has advocated for clients throughout his career in the world of telecommunications, specifically pertaining to technical and regulatory issues. He has taught business management and marketing as an adjunct professor at colleges in the St. Louis area for over 15 years.

Bart is a graduate of the University of Kansas and has a M.B.A. from Washington University in St. Louis.

As the Information Systems Project Manager on major projects such as Equal Access, and later in key roles interfacing with AT&T, MCI, Sprint, and the Department of Justice, Bart has delivered software solutions and consultation with business leaders. In addition, he has worked with startup CLECs and public safety agencies to solve communications issues, ultimately saving lives, and he served as the Project Manager for the CSI and for the NINGA NG9-1-1 projects. In his spare time, Bart enjoys fly fishing and spending time with his growing cohort of grandchildren.

Index

F

Recent Titles in the Artech House Mobile Communications Series

William Webb, Series Editor

3G CDMA2000 Wireless System Engineering, Samuel C. Yang

3G Multimedia Network Services, Accounting, and User Profiles, Freddy Ghys, Marcel Mampaey, Michel Smouts, and Arto Vaaraniemi

5G New Radio: Beyond Mobile Broadband, Amitav Mukherjee

5G Spectrum and Standards, Geoff Varrall

802.11 WLANs and IP Networking: Security, QoS, and Mobility, Anand R. Prasad and Neeli R. Prasad

Achieving Interoperability in Critical IT and Communications Systems, Robert I. Desourdis, Peter J. Rosamilia, Christopher P. Jacobson, James E. Sinclair, and James R. McClure

Advances in 3G Enhanced Technologies for Wireless Communications, Jiangzhou Wang and Tung-Sang Ng, editors

Advances in Mobile Information Systems, John Walker, editor

Advances in Mobile Radio Access Networks, Y. Jay Guo

Applied Satellite Navigation Using GPS, GALILEO, and Augmentation Systems, Ramjee Prasad and Marina Ruggieri

Artificial Intelligence in Wireless Communications, Thomas W. Rondeau and Charles W. Bostian

Broadband Wireless Access and Local Network: Mobile WiMax and WiFi, Byeong Gi Lee and Sunghyun Choi

CDMA for Wireless Personal Communications, Ramjee Prasad

CDMA Mobile Radio Design, John B. Groe and Lawrence E. Larson

CDMA RF System Engineering, Samuel C. Yang

CDMA Systems Capacity Engineering, Kiseon Kim and Insoo Koo

CDMA Systems Engineering Handbook, Jhong S. Lee and Leonard E. Miller

Cell Planning for Wireless Communications, Manuel F. Cátedra and Jesús Pérez-Arriaga

Cellular Communications: Worldwide Market Development, Garry A. Garrard

Cellular Mobile Systems Engineering, Saleh Faruque

Cognitive Radio Interoperability through Waveform Reconfiguration, Leszek Lechowicz and Mieczyslaw M. Kokar

Cognitive Radio Techniques: Spectrum Sensing, Interference Mitigation, and Localization, Kandeepan Sithamparanathan and Andrea Giorgetti

The Complete Wireless Communications Professional: A Guide for Engineers and Managers, William Webb

Digital Communication Systems Engineering with Software-Defined Radio, Di Pu and Alexander M. Wyglinski

EDGE for Mobile Internet, Emmanuel Seurre, Patrick Savelli, and Pierre-Jean Pietri

Emerging Public Safety Wireless Communication Systems, Robert I. Desourdis, Jr., et al.

From LTE to LTE-Advanced Pro and 5G, Moe Rahnema and Marcin Dryjanski

The Future of Wireless Communications, William Webb

Geographic Information Systems Demystified, Stephen R. Galati

Geospatial Computing in Mobile Devices, Ruizhi Chen and Robert Guinness

GPRS for Mobile Internet, Emmanuel Seurre, Patrick Savelli, and Pierre-Jean Pietri

GPRS: Gateway to Third Generation Mobile Networks, Gunnar Heine and Holger Sagkob

GSM and Personal Communications Handbook, Siegmund M. Redl, Matthias K. Weber, and Malcolm W. Oliphant

GSM Networks: Protocols, Terminology, and Implementation, Gunnar Heine

GSM System Engineering, Asha Mehrotra

Handbook of Land-Mobile Radio System Coverage, Garry C. Hess

Handbook of Mobile Radio Networks, Sami Tabbane

Handbook of Next-Generation Emergency Services, Barbara Kemp and Bart Lovett

High-Speed Wireless ATM and LANs, Benny Bing

Implementing Full Duplexing for 5G, David B. Cruickshank

Inside Bluetooth Low Energy, Second Edition, Naresh Gupta

Interference Analysis and Reduction for Wireless Systems, Peter Stavroulakis

Interference and Resource Management in Heterogeneous Wireless Networks, Jiandong Li, Min Sheng, Xijun Wang, and Hongguang Sun

Internet Technologies for Fixed and Mobile Networks, Toni Janevski

Introduction to 3G Mobile Communications, Second Edition, Juha Korhonen

Introduction to 4G Mobile Communications, Juha Korhonen

Introduction to Communication Systems Simulation, Maurice Schiff

Introduction to Digital Professional Mobile Radio, Hans-Peter A. Ketterling

Introduction to GPS: The Global Positioning System, Ahmed El-Rabbany

An Introduction to GSM, Siegmund M. Redl, Matthias K. Weber, and Malcolm W. Oliphant

Introduction to Mobile Communications Engineering, José M. Hernando and F. Pérez-Fontán

Introduction to OFDM Receiver Design and Simulation, Y. J. Liu

Introduction to Radio Propagation for Fixed and Mobile Communications, John Doble

Introduction to Wireless Local Loop, Broadband and Narrowband, Systems, Second Edition, William Webb

IS-136 TDMA Technology, Economics, and Services, Lawrence Harte, Adrian Smith, and Charles A. Jacobs

Location Management and Routing in Mobile Wireless Networks, Amitava Mukherjee, Somprakash Bandyopadhyay, and Debashis Saha

LTE Air Interface Protocols, Mohammad T. Kawser

Metro Ethernet Services for LTE Backhaul, Roman Krzanowski

Mobile Data Communications Systems, Peter Wong and David Britland

Mobile IP Technology for M-Business, Mark Norris

Mobile Satellite Communications, Shingo Ohmori, Hiromitsu Wakana, and Seiichiro Kawase

Mobile Telecommunications Standards: GSM, UMTS, TETRA, and ERMES, Rudi Bekkers

Mobile-to-Mobile Wireless Channels, Alenka Zajić

Mobile Telecommunications: Standards, Regulation, and Applications, Rudi Bekkers and Jan Smits

Multiantenna Digital Radio Transmission, Massimiliano Martone

Multiantenna Wireless Communications Systems, Sergio Barbarossa

Multi-Gigabit Microwave and Millimeter-Wave Wireless Communications, Jonathan Wells

Multipath Phenomena in Cellular Networks, Nathan Blaunstein and Jørgen Bach Andersen

Multiuser Detection in CDMA Mobile Terminals, Piero Castoldi

OFDMA for Broadband Wireless Access, Slawomir Pietrzyk

Personal Wireless Communication with DECT and PWT, John Phillips and Gerard MacNamee

Practical Wireless Data Modem Design, Jonathon Y. C. Cheah

The Practitioner's Guide to Cellular IoT, Cameron Kelly Coursey

Prime Codes with Applications to CDMA Optical and Wireless Networks, Guu-Chang Yang and Wing C. Kwong

Quantitative Analysis of Cognitive Radio and Network Performance, Preston Marshall

QoS in Integrated 3G Networks, Robert Lloyd-Evans

Radio Engineering for Wireless Communication and Sensor Applications, Antti V. Räisänen and Arto Lehto

Radio Propagation in Cellular Networks, Nathan Blaunstein

Radio Resource Management for Wireless Networks, Jens Zander and Seong-Lyun Kim

Radiowave Propagation and Antennas for Personal Communications, Third Edition, Kazimierz Siwiak and Yasaman Bahreini

RDS: The Radio Data System, Dietmar Kopitz and Bev Marks

Resource Allocation in Hierarchical Cellular Systems, Lauro Ortigoza-Guerrero and A. Hamid Aghvami

RF and Baseband Techniques for Software-Defined Radio, Peter B. Kenington

RF and Microwave Circuit Design for Wireless Communications, Lawrence E. Larson, editor

RF Positioning: Fundamentals, Applications, and Tools, Rafael Saraiva Campos and Lisandro Lovisolo

Sample Rate Conversion in Software Configurable Radios, Tim Hentschel

Signal Processing Applications in CDMA Communications, Hui Liu

Signal Processing for RF Circuit Impairment Mitigation, Xinping Huang, Zhiwen Zhu, and Henry Leung

Smart Antenna Engineering, Ahmed El Zooghby

Software-Defined Radio for Engineers, Travis F. Collins, Robin Getz, Di Pu, and Alexander M. Wyglinski

Software Defined Radio for 3G, Paul Burns

Spread Spectrum CDMA Systems for Wireless Communications, Savo G. Glisic and Branka Vucetic

Technical Foundations of the IoT, Boris Adryan, Dominik Obermaier, and Paul Fremantle

Technologies and Systems for Access and Transport Networks, Jan A. Audestad

Third-Generation and Wideband HF Radio Communications, Eric E. Johnson, Eric Koski, William N. Furman, Mark Jorgenson, and John Nieto

Third Generation Wireless Systems, Volume 1: Post-Shannon Signal Architectures, George M. Calhoun

Traffic Analysis and Design of Wireless IP Networks, Toni Janevski

Transmission Systems Design Handbook for Wireless Networks, Harvey Lehpamer

UMTS and Mobile Computing, Alexander Joseph Huber and Josef Franz Huber

Understanding Cellular Radio, William Webb

Understanding Digital PCS: The TDMA Standard, Cameron Kelly Coursey

Understanding GPS: Principles and Applications, Second Edition, Elliott D. Kaplan and Christopher J. Hegarty, editors

Understanding WAP: Wireless Applications, Devices, and Services, Marcel van der Heijden and Marcus Taylor, editors

Universal Wireless Personal Communications, Ramjee Prasad

WCDMA: Towards IP Mobility and Mobile Internet, Tero Ojanperä and Ramjee Prasad, editors

Wireless Communications in Developing Countries: Cellular and Satellite Systems, Rachael E. Schwartz

Wireless Communications Evolution to 3G and Beyond, Saad Z. Asif

Wireless Intelligent Networking, Gerry Christensen, Paul G. Florack and Robert Duncan

Wireless LAN Standards and Applications, Asunción Santamaría and Francisco J. López-Hernández, editors

Wireless Sensor and Ad Hoc Networks Under Diversified Network Scenarios, Subir Kumar Sarkar

Wireless Technician's Handbook, Second Edition, Andrew Miceli

For further information on these and other Artech House titles,including previously considered out-of-print books now available through our In-Print-Forever® (IPF®) program, contact:

Artech House
685 Canton Street
Norwood, MA 02062
Phone: 781-769-9750
Fax: 781 769-6334
e-mail: artech@artechhouse.com

Artech House
16 Sussex Street
London SW1V 4RW UK
Phone: +44 (0)20 7596-8750
Fax: +44 (0)20 7630-0166
e-mail: artech-uk@artechhouse.com

Find us on the World Wide Web at: www.artechhouse.com